An Introduction to Hadronic Mechanics, Volume I

An Introduction to Hadronic Mechanics, Volume I:

Applications to Nuclear Physics

By

Ruggero Maria Santilli

Cambridge Scholars Publishing

An Introduction to Hadronic Mechanics, Volume I:
Applications to Nuclear Physics

By Ruggero Maria Santilli

This book first published 2026

Cambridge Scholars Publishing

Lady Stephenson Library, Newcastle upon Tyne, NE6 2PA, UK

British Library Cataloguing in Publication Data
A catalogue record for this book is available from the British Library

ISBN: 978-1-0364-7004-3
ISBN (Ebook): 978-1-0364-7005-0

This book is dedicated to my friend

Professor Erik Trell, M. D.
University of Linköping, Sweden

for his uniquely eclectic knowledge of
mathematics, physics, chemistry and biology

TABLE OF CONTENTS

PREFACE

Since the early 20th century, the study of physics has long been dominated by the principles of quantum mechanics including Heisenberg's uncertainty principle and Pauli exclusion principle. While these theories have provided significant insights into the behaviour of particles at the atomic level, they have also encountered limitations at deeper structural levels as first indicated by A. Einstein, B. Podolsky and N. Rosen (EPR) with their 1934 historical view that *"Quantum mechanics is not a complete theory"* [1] (Chapter 1).

By using a language accessible to the general educated audience, this book *"An Introduction to Hadronic Mechanics, Volume I: Applications to Nuclear Physics*" attempts to encapsulate the mathematical, theoretical, experimental and industrial advancements made to date in the EPR completion of quantum mechanics into a new discipline for strongly interacting particles, that I therefore called *hadronic mechanic* [3] (see also subsequent monographs [4]-[10]; collection of recent independent papers [11]; recent applications in nuclear physics [12]-[17], chemistry [18] [19] and biology [20] [21]; and recent reviews [22]-[26]).

I initiated the studies with my 1965 Ph. D. Thesis in Theoretical Physics at the University of Torino, Italy, and continued from September 1977 to August 1981 at Harvard University under DOE support and from September 1981 to date at the Institute for Basic Research, which studies have been complemented by numerous contributions from mathematicians, theoreticians and experimentalists with applications in all quantitative sciences.

The first chapters introduce fundamental concepts on the need for the EPR completion of quantum into hadronic mechanics, providing an overview of its development within a historical context. We delve into the experimental foundations supporting the theory, highlighting key evidence and observations demonstrating that the extended charge distributions of protons and neutrons are in conditions of deep mutual penetration and entanglement, resulting in the presence of contact/non-Hamiltonian interactions [2] which are outside the representational capabilities of quantum mechanics, yet represented in hadronic mechanics via the completion of all associative products AB of quantum mechanics into the associativity-preserving form A⋆B = ASB, where S is a positive-definite companion of the Hamiltonian H [3], nowadays called the Santillian [21-[25], also representing Bohm's hidden variables as being hidden in the axiom of associativity.

Despite its simplicity, the representation of non-Hamiltonian interactions with the completed product A⋆B = ASB requires, for consistency, the corresponding completion of the multiplicative unit, from the equally millenary number 1 of quantum mechanics into the positive-definite quantity $1^* = 1/S > 0$. In turn, the indicated axiom-preserving completions of the multiplicative unit and related product requires, when applicable, a corresponding compatible re-formulation of the totality of 20th century applied mathematics and related quantitative sciences with no exception known to me [25], thus explaining the decades of efforts that resulted to be needed in the achievement of maturity in the completion of quantum into hadronic mechanics.

Subsequent chapters trace the origination and evolution of hadronic mechanics, detailing my early studies and the theoretical advancements that led to the development of the new Lie-admissible and Lie-isotopic mathematics and related branches of hadronic mechanics to represent time irreversible (Lie-admissible) and reversible (Lie-isotopic) bound states of extended charge distributions/wave packets in condition of mutual penetration with potential Hamiltonian and contact non-Hamiltonian internal forces as occurring in nuclear, molecular or biological structures.

Applications and advancements in nuclear physics [12]-[17] form a crucial part of this first volume. Following the study of the new mathematical and physical methods needed for the representation of extended wave packets in condition of deep mutual entanglement, we review in Chapter 8 the achievement by hadronic mechanics of the first known numerically exact and time invariant representation of the synthesis of the neutron from a proton and an electron in the core of star, as well as the Deuteron and other stable nuclei (Sections 8.2 to 8.6). We then review the extensive tests and certifications of the excess clean energy produced by Intermediate Controlled Nuclear Fusions (ICNF) of light natural elements, (Section 8.7). We finally review the sustained and controlled synthesis of negatively charged pseudo-nuclei (hadronic bonds of electron pairs in singlet coupling and natural nuclei thanks to their extremely big Coulomb attraction) which are solely possible under the hadronic uncertainty principle [12] [24], and their natural attraction in singlet coupling by ordinary positively charged nuclei, resulting in the inevitable HyperFusions with the production of clean nuclear energy without the Coulomb barrier and without harmful radiation or waste (Section 8.8).

We then review the experimental and industrial efforts to achieve the novel HyperFusions which are sustainable nuclear fusions of light natural elements without the Coulomb barrier and without the emission of harmful radiation. (Section 9.8).

By recalling that Heisenberg's uncertainty principle is experimentally established solely for point like particles in vacuum under electromagnetic and weak interactions, particular attention is dedicated to the fact that the above advances are solely possible under the progressive weakening of Eisenberg's uncertainty principle for extended particles in deep mutual entanglement in the structure of hadrons, nuclei and stars, with the full recovering of Einstein's determinism at the Schwartzschild's horizon for black holes [12] [25] (Chapter 6).

The most important advancement by the EPR completion of quantum chemistry into *hadronic chemistry* [7] has been the first known identification of the attraction between the identical electrons of valence bonds [7] with ensuing quantitative representations of molecular structures, resulting in the first known numerically exact representation of the experimental data of the Hydrogen [18] and water molecules [19].

In a future volume, I plan to review the most important advancements achieved by the EPR completion of quantum biology into *hadronic biology* [8], such as the first known mathematical representation of life intended as the difference between organic and inorganic molecules [20] with ensuing first known quantitative representation of biological events or behavior that are generally considered to be supernatural [21] (see volumes [15] for independent contributions).

The experimental verifications and case studies presented here provide concrete evidence of the validity and applicability of hadronic mechanics. These studies affirm the theoretical constructs and illustrate their potential to address unresolved problems in physics, chemistry and biology.

Readers exposed for the first time to the new mathematics and mechanics are suggested to read first the overview presented in the Executive Summary of Sect. 9, or see my 30' lecture at the 2025 Materials 2.0, Cambridge, U. K., https://www.world-lecture-series.org/level-XVIII and then pass to the reading of preferred topics

Most of the original insights on hadronic mechanics were first published in the Hadronic Journal, which was founded for that purpose in April 1978 at Harvard University by Profs. H. Georgi and R. M. Santilli with the Editorship of numerous distinguished scholars around the world under the administration of my wife Carla Santilli as President of Hadronic Press, Inc. www.hadronicpress.com. Since that time, the Hadronic Journal has continued regular production for forty-seven years by publishing at no cost a large number of refereed papers, Ph. D. theses, proceedings of workshops, and international conferences on hadronic mechanics cited in this book for interested historians.

As we conclude, we reflect on the future directions of this evolving field. The book highlights open problems, potential advancements, and areas ripe for further research. This book is intended for scientists and students keen on exploring the frontiers of physics beyond the traditional paradigms. I hope this work will inspire further research and foster a deeper understanding of the complex interactions that govern the behaviour of matter at the most fundamental level.

I extend my deepest gratitude to all colleagues who have contributed to the development of hadronic mechanics and to my readers, who continue to seek knowledge and understanding in this ever-expanding field of study with particular reference to Professors H. Ahmar, A.O. E. Animalu, A. K. Aringazin, A. Bayoumi, S. Beghella-Bartoli, T. Bhadra Man, A. Bhalekar, R. Brenna, C. Burande, W. Cai, P. Caldirola, I. B. Das Sarma, B. Davvaz, S. S. Dhondge, J. Dunning-Davies, I. Gandzha, R. M. F. Ganfornina, S. Georgiev, T. Gill, V. de Haan, C.-X. Jiang, A. Jannussis, . E. Johansen, J. V. Kadeisvili, T. Kuliczkowski, J. Lohmus, R. Mignani, A. P. Mills, R. Miron, R. Perez-Enriquez, M. R. Molaei, A. Muktibodh, H. C. Myung, A. A. Nassikas, M. Nishioka, R. Norman, Z. Oziewicz, J. Rak, E. Recami, A. Shoeber, D. S. Sourlas, J. N. Valdez, E. Trell, B. Veljanoski, Gr. T. Tsagas, T. Vougiouklis, H. E. Wilhelm, Y. Yang, L. Ying, and others.

I would like to express my sincere thanks to: Professors N. Muktibodh from India and S. Georgiev from France for a review of the mathematical content of the book; Professors A. K. Aringazin from Kazakhstan, J. Dunning Davies from England, F. Inglese from Italy, Lessner from Israel and E. Velardo from Italy for a review of the physical content of the book; S. Beghella Bartoli from Italy and M. Kashchenko from Russia for a control of the industrial applications of the book; Mrs. Sherri Stone, of the Hadronic Press, for a linguistic control; and my Artificial Intelligence for technical control of the book and for composing the two indeces

Ruggero Maria Santilli
October 15, 2025
research@i-b-r.org

www.i-b-r.org/Sir-Santilli-bionotes-05-15-21.pdf

https://www.i-b-r.org/Ruggero-Maria-Santilli.htm

References

[1] Einstein, A. Podolsky, B. and Rosen, N.: Can quantum-mechanical description of physical reality be considered complete? Physical Review Vol. 47, 777-780 (1935), http://www.eprdebates.org/docs/epr-argument.pdf

[2] Santilli, R. M.: *Foundation of* Theoretical *Mechanics,* Springer-Verlag, Heidelberg, Germany, Vol. I (1978) T*he Inverse Problem in Newtonian Mechanics*, http://www.santilli-foundation.org/docs/Santilli-209.pdf

[3] Santilli, R. M.: *Foundation of Theoretical Mechanics*, Springer-Verlag, Heidelberg, Germany, Vol. II (1983) *Birkhoffian Generalization of Hamiltonian Mechanics,* www.santilli-foundation.org/docs/santilli-69.pdf

[4] Santilli, R. M.: *Elements of Hadronic Mechanics,* Ukraine Academy of Sciences, Kiev (1995), Vol. I, *Mathematical Foundations,* www.santilli-foundation.org/docs/Santilli-300.pdf

[5] Santilli, R. M.: *Elements of Hadronic Mechanics,* Ukraine Academy of Sciences, Kiev (1995), Vol. II, *Theoretical Foundations,* www.santilli-foundation.org/docs/Santilli-301.pdf

[6] Santilli, R. M.: *Elements of Hadronic Mechanics,* Ukraine Academy of Sciences, Kiev (2016), Vol. III: *Experimental* verifications, www.santilli-foundation.org/docs/elements-hadronic-mechanics-iii.compressed.pdf

[7] Santilli, R. M.: *Foundations of Hadronic Chemistry with Applications to New Clean Energies and Fuels,* Kluwer Academic Publishers (2001), http://www.santilli-foundation.org/docs/Santilli-113.pdf Russian translation by Aringazin, A. K. http://i-b-r.org/docs/Santilli-Hadronic-Chemistry.pdf

[8] Santilli, R. M: *Isotopic, Genotopic and Hyperstructural Methods in Theoretical Biology,* Ukraine Academy of Sciences, Kiev (1994), http://www.santilli-foundation.org/docs/santilli-67.pdf

[9] Santilli, R. M.: *Isotopic Generalizations of Galilei and Einstein Relativities,* International Academic Press (1991), Vol. I: *Mathematical Foundations,* http://www.santilli-foundation.org/docs/Santilli-01.pdf

[10] Santilli, R. M.: *Isotopic Generalizations of Galilei and Einstein Relativities*, International Academic Press (1991), Vol. II: Classical Formulations, www.santilli-foundation.org/docs/Santilli-61.pdf

[11] Anderson R., Editor: *Hadronic Mathematics, Mechanics and Chemistry,* Volumes I, II, III, International Academic Press, (2008), http://www.i-b-r.org/Hadronic-Mechanics.htm

[12] Santilli, R. M.: Studies on the classical determinism predicted by A. Einstein, B. Podolsky and N. Rosen, Ratio Mathematica Vol. 37, 5-23 (2019), www.eprdebates.org/docs/epr-paper-ii.pdf

[13] Santilli, R. M.: Lie-isotopic representation of stable nuclei I: Apparent insufficiencies of quantum mechanics in nuclear physics Ratio Mathematica Vol. 52, 43-63 (2024), https://www.santilli-foundation.org/docs/3-RM-SI-Vol-52-2024-fin.pdf

[14] Santilli, R. M.: Lie-isotopic representation of stable nuclei II: Exact and time invariant representation of the Deuteron data, Ratio Mathematica Vol. 52, 64-130 (2024), https://www.santilli-foundation.org/docs/4-RM-SI-Vol-52-2024-fin.pdf

[15] Santilli, R. M.: Lie-isotopic representation of stable nuclei III: Exact and time invariant representation of nuclear stability Ratio Mathematica Vol. 52, 131-175 (2024), https://www.santilli-foundation.org/docs/5-RM-SI-Vol-52-2024-fin.pdf

[16] Santilli, R. M.: Reduction of Matter in the Universe to Protons and Electrons via the Lie-isotopic Branch of Hadronic Mechanics, Progress in Physics, Vol. 19, 73-99 (2023), https://www.ptep-online.com/2023/PP-65-09.PDF

[17] Santilli, R. M.: Apparent Resolution of the Coulomb Barrier for Nuclear Fusions Via the Irreversible Lie-admissible Branch of Hadronic Mechanics, Progress in Physics Vol. 18, 138-163 (2022), ttps://www.santilli-foundation.org/hyperfusion-2022.pdf

[18] Santilli, R. M. and Shillady, D. D.: A new isochemical model of the hydrogen molecule, Intern. J. Hydrogen Energy Vol. 24, 943-951 (1999), http://www.santilli- foundation.org/docs/Santilli-135.pdf

[19] Santilli, R. M. and Shillady, D. D.: A new isochemical model of the water molecule, Intern. J. Hydrogen Energy Vol. 25, 173-196 (2000), http://www.santilli-foundation.org/docs/Santilli-39.pdf

[20] Santilli R. M. and T. Vougiouklis, A New Conception of Living Organisms and its Representation via Lie-Admissible Hv-Hyperstructures, Algebras, Groups and Geometries Vol. 37, 741-764 (2020), http://www.santilli-foundation.org/docs/Santilli-Vougiouklis-2020-epr.pdf

[21] Inglese F, Velardo E., Santilli R. M. and Vougiouklis T., Lie-admissible irreversible biological entanglements and their apparent initiation of hadronic medicine Ratio Mathematica Vol. 54, 1-35 (2025), http://www.santilli-foundation.org/docs/hadronic-medicine.pdf

[22] Editors: Progressive Recovering of Einstein's Determinism under Strong Interactions, Scientia November 2024, https://doi.org/10.33548/SCIENTIA1115

[23] Editors: Completing Quantum Mechanics: Hadronic Mechanics and its Potential for Clean Energy, Scientia December 2024, https://doi.org/10.33548/SCIENTIA1133

[24] Muktibodh, A: Santilli's recovering of Einstein's determinism, Progress in Physics 20, 26-34 (2024), http://www.eprdebates.org/docs/muktibodh-2024.pdf

[25] Velardo, E: An introduction to iso-, geno- and hyper- mathematics and their applications in physics, chemistry and biology, Ratio Mathematica, Vol. 54, 118- 166 (2025), http://www.santilli-foundation.org/docs/Velardo-2025.pdf

CHAPTER 1

APPARENT INSUFFICIENCIES OF QUANTUM MECHANICS IN NUCLEAR PHYSICS

1.1. Introduction

Quantum mechanics, with its probabilistic nature and wave-particle duality, has profoundly increased our understanding of the microscopic world. Despite its successes in the exact representation of point-like particles in a vacuum under electromagnetic interactions, notable physicists, including Einstein, have questioned the completeness of quantum mechanics, arguing that it fails to represent the entirety of physical reality. By using a language accessible to the general physics audience, this chapter explores Albert Einstein's [1-1] critique of quantum mechanics, Niels Bohr's [1-2] rebuttal, and R. M. Santilli's [1-3] identification of apparent insufficiencies of quantum mechanics in nuclear physics (see also Refs. [1-4] [1-5] and vast literature quoted therein).

1.2. Einstein's Critique of Quantum Mechanics

In 1935, Albert Einstein and hid graduate students Boris Podolsky and Nathan Rosen (EPR) [1-1] published a seminal paper titled "Can Quantum-Mechanical Description of Physical Reality Be Complete?" In this paper, they argued that quantum mechanics does not entirely describe reality. They introduced the concept of "elements of reality," suggesting that if a physical quantity can be predicted with certainty without disturbing the system, it must correspond to an element of reality. According to EPR, Quantum mechanics fails to account for such elements fully.

Einstein's discomfort with quantum mechanics stemmed from its inherent indeterminism and the inability of quantum mechanical wave packets to represent the entire physical reality. He believed that quantum mechanics should eventually be supplemented by a more comprehensive theory that could restore classical determinism under some limit conditions.

The EPR paper presented a thought experiment involving two particles that interacted and separated. According to quantum mechanics, the particles become entangled, meaning the measurement of one particle's state instantaneously determines the state of the other, regardless of the distance between them. This phenomenon, which Einstein famously referred to as "spooky action at a distance," seemed to violate the principle of locality.

EPR argued that if quantum mechanics were complete, it should be possible to determine both particles' position and momentum simultaneously. The paradox they presented showed that either quantum mechanics is incomplete or requires abandoning the locality principle.

1.3. Bohr's Rebuttal

Niels Bohr, [1-2] a staunch defender of quantum mechanics, responded to the EPR paper by emphasizing the role of measurement in defining physical reality. According to Bohr's Copenhagen interpretation, quantum mechanics does not describe an objective reality independent of observation. Instead, it provides probabilities for different measurement outcomes.

Bohr argued that the EPR thought experiment did not reveal any incompleteness in quantum mechanics but instead highlighted the fundamental limitations of classical concepts when applied to quantum phenomena. He maintained that measurement disturbs the system, making it impossible to attribute definite values to physical quantities before measurement.

Bohr argued that the EPR thought experiment did not reveal any incompleteness in quantum mechanics but instead highlighted the fundamental limitations of classical concepts when applied to quantum phenomena. He maintained that measurement disturbs the system, making it impossible to attribute definite values to physical quantities before measurements.

1.4. Apparent Experimental Insufficiencies

In support of the EPR argument [1-1], Ruggero Maria Santilli [1-3] argues that non-relativistic and relativistic quantum mechanics are not complete theories, firstly, because of several apparent insufficiencies in the representation of nuclear experimental data and, secondly, because of apparent axiomatic insufficiencies for the representation of complex structures in the universe.

1.4-1. Inability to represent the synthesis of the neutron from the Hydrogen atom in the core of stars

Stars initiate their lives as an aggregate of Hydrogen and must synthesize the neutron from the Hydrogen as a condition to synthesize the Deuteron from a proton and a neutron, and finally, initiate the production of light via the synthesis of the Helium from two Deuterons. Despite the extremely large Coulomb attraction between the electron and the proton due to their opposite charges, quantum mechanics cannot represent the synthesis of the neutron from a proton and an electron because prohibited by Heisenberg's uncertainty principle since the standard deviation of the coordinates of the electron is a large multiple of the size of the neutron. The standard deviation of its linear momentum implies the electron energy is a multiple of the neutron's. Additionally, the synthesis of the neutron from a proton and an electron requires energy because the mass of the neutron is 0.782 MeV *bigger* than the sum of the masses of the proton and of the electron, thus requiring a "mass excess," which is anathema for quantum mechanics due to its sole possibility of representing the known "mass defect" in nuclear fusions. (Sect. 2.3, p. 47 of [1-3]).

1.4-2. Inability to achieve an exact representation of nuclear magnetic moments

The quantum mechanical computation of the magnetic moment of the smallest possible atom, the Deuteron, from those of the nuclear constituents, the proton and the neutron, is about 3% off the experimental value, with deviations for heavier nuclei, such as the Zirconium, so large to prevent a scientific analysis (Sect. 2.6, p. 50 of [1-3] and Fig. 1.1).

1.4-3. Inability to achieve a consistent representation of nuclear spins

The sole stable bound state predicted by quantum mechanics for spinning particles is that they couple with opposite spins. In contrast, the coupling with parallel spins is highly unstable, as for ordinary gears. The application of the above basic law to the spin of the Deuteron is that the proton and the neutron should couple with opposite spins, in which case the total angular momentum of the Deuteron is zero in gross disagreement with experimental measurements for which said spin is one. When facing such an inconsistency, mainstream nuclear physicists claim that the total value one is due to orbital contributions, which is, again, in gross disagreement with experiments since the total spin one is measured for the Deuteron in its ground state, which by definition, is the state with null orbital contributions. In passing to the spin of heavier nuclei, the deviations between quantum mechanical predictions and the experimental values of nuclear spins are too large to prevent a scientific analysis (Sect. 2.5, p. 49 of [1-3] and Fig. 1.2).

1.4-4. Inability to represent nuclear stability despite the neutron's natural instability

Following its synthesis from a proton and an electron, the neutron is naturally unstable since it decays spontaneously into its original constituents in about 15 minutes. However, all organic and inorganic matter in our environment contains stable nuclei that, according to mainstream nuclear physicists, are composed of protons and neutrons. This visual evidence implies the transition of the neutron from an unstable state in a vacuum to a permanently stable state in a nuclear structure, which is prohibited by the special relativity of relativistic quantum mechanics, technically, by the invariance of masses under the fundamental Poincare' symmetry (Sect. 2.4, p. 59 of [1-3]).

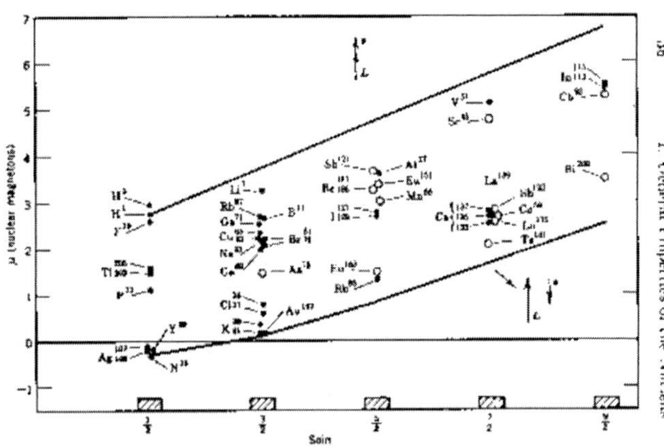

Figure 1.1. A historical insufficiency of quantum mechanics in nuclear physics is the inability to achieve in one century an exact representation of nuclear magnetic moments, with 3% deviations for the Deuteron and embarrassing deviations for heavy nuclei reported in this figure.

1.4-5. Inability to represent nuclear stability despite the strongly repulsive protonic Coulomb force

Neutrons have no electric charge, while protons are positively charged. Consequently, the protons of permanently stable nuclei in our environment experience a repulsive Coulomb force which, according to simple calculations, is of the order of thousands of Newtons, thus being so astronomical in value for nuclear standards that available quantum mechanical models of nuclear forces cannot overcome. It follows that, after one century of research, quantum mechanics has been unable to represent nuclear stability (Sect. 2.4, p. 59 of [1-3]).

1.4-6. Inability to predict industrially viable recycling of radioactive nuclear waste

In Santilli's view, one of the potential historical problems of our decaying environment is that quantum mechanics has been unable to predict, let alone develop, any industrially viable means for the recycling of radioactive nuclear waste, that consequently, had to be stored by nuclear power plants in their facilities since the beginning of their operations up to the current alarming storage levels. Given the known opposition against the transportation of waste in public roads, the only industrially effective means is recycling radioactive nuclear waste by the nuclear power plants themselves in their own facilities. In turn, such recycling can only be done via nuclear transmutations of very long into very short mean lives that non-relativistic and relativistic quantum mechanics strictly prohibit. [1-3]

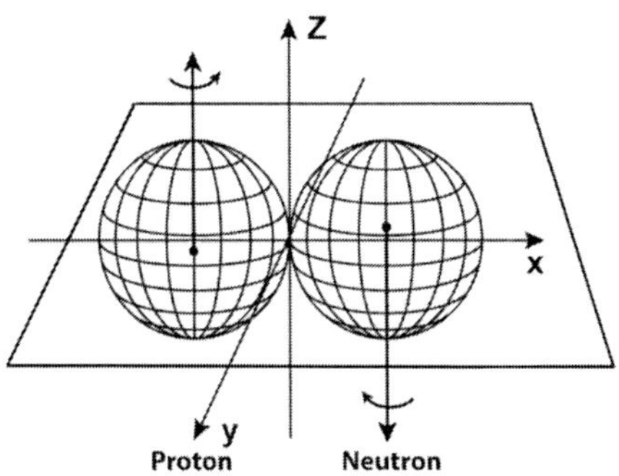

Figure 1.2. Quantum mechanics cannot represent the spin S = 1 of the Deuteron because the sole possible stable bound state between a proton and a neutron both with spin 1/2 is the singlet with spin S = 0 depicted in this figure, by therefore

requiring the century old, hypothetical assumption of orbital states with angular momentum L = 1 against the measurement of the Deuteron spin S = 1 in its ground state. Greater departures between quantum mechanical predictions and experimental evidence occur for the spin of heavier nuclei.

1.4-7. Inability to represent controlled and sustainable clean nuclear fusions

In Santilli's view, the biggest, potentially historical reason for the alarming decay of our environment is the century-old inability of non-relativistic and relativistic quantum mechanics to achieve sustainable nuclear fusions. A first reason for this insufficiency is the known inability by quantum mechanics to represent the irreversibility of nuclear fusions due to its strictly conservative axiomatic structure. A second reason is the inability to bypass the Coulomb barrier, i.e., the repulsive Coulomb force between natural, positively charged nuclei, which has to be overcome to achieve nuclear contact. Nuclear fusions are unavoidable at this point due to the activation of the strongly attractive nuclear force. The potentially historical character of these insufficiencies is the vastly ignored evidence that the repulsive force of the Coulomb barrier has such an astronomical value indicated earlier to prevent a plausible overcome by mechanical, magnetic, and other conventional means, while much needed. Heisenberg's uncertainty principle generally prohibits new approaches.

1.5. Apparent axiomatic insufficiencies

Santilli continues the analysis in paper [1-3] by identifying the apparent axiomatic origin of insufficiencies 1.4-1 to 1.4-7, whose solution is attempted in Chapter 3 for possible completion of quantum mechanics according to the EPR argument [1-1].

1.5-1. Locality

Non-relativistic and relativistic quantum mechanics, including their Galilean and Lorentz symmetries and ensuing relativities, are local theories [1-1] because they can solely be defined at one point in space. Such an axiomatic structure has been experimentally proved valid for point-like particles in a vacuum under long-range electromagnetic and other interactions. This is true for atomic electrons, particles in accelerators, crystals, and others. However, the same locality is expected to be insufficient for nuclear structures, complex chemical compounds, and numerous other cases in the universe in which we have extended constituents in conditions of mutual penetration, resulting in basically new contact, thus zero-range interactions that are extended over volumes not reducible to a finite set of isolated points, thus requiring a non-local completion of quantum mechanics. Santilli argues that the mathematically, physically, and chemically consistent treatment of dimensionless particles requires local theories. However, the representation of the physical or chemical reality of the extended character of particles implies, per se, the inapplicability (rather than the violation) of local theories in favor of their non-local completion [1-1] under the condition of unambiguously recovering conventional local theories for mutual distances sufficiently large to permit the point-like abstraction of physical particles (Sect. 3.2, p. 53 of [1-3]).

1.5-2. Linearity

The dynamical equations of quantum mechanics, such as Schroedinger's equation, are linear in the sense of depending on the first power of the wave function. However, there are several technical arguments, e.g., to represent the entanglement of particles, suggesting the need for the non-linear completion of linear theories [1-1], i.e., theories with dynamical equations depending on arbitrary powers of the wave function as a condition to represent broader physical realities (Sect. 3.1, p. 52 of [1-3]).

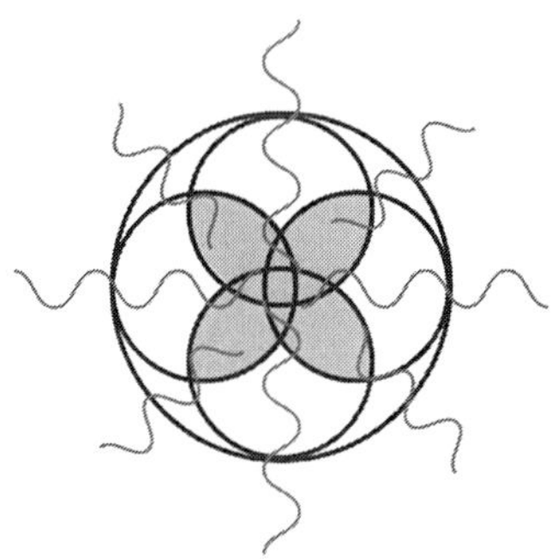

Figure 1.3. According to half a century of measurements, the charge distribution of the proton and the neutron has the radius $R = 0.87 \times 10^{-13}$ cm while the charge distribution of the Helium has the radius $R = 1.6 \times 10^{-13}$ cm. Consequently, the two protons and the two neutrons in the Helium structure are in condition of partial mutual penetration of their hyperdense structures for 0.07 \times 10^{-13} cm, with ensuing non-linear, non-local and non-potential interactions beyond any hope of representation via quantum mechanics.

1.5-3. Potentiality

The abstraction of particles as being point-like automatically restricts all possible interactions to be derivable from a potential. By contrast, the representation of the actual physical dimension of particles leaves unchanged potential interactions due to their long-range character yet requires their completion with contact zero-range interactions that, as such, are not derivable from a potential. Santilli's main argument outlined in this book is that a most important completion of quantum mechanics according to the EPR argument [1-1] appears to be that with a mathematically, physically, and chemically consistent representation of non-potential interactions since they automatically imply the quantitative representation of the extended character of particles and their non-local and non-linear interactions (Sect. 3.3, p. 54 of [1-3]).

1.5-4. Time reversibility

In Santilli's [1-3] view, the most serious limitation of quantum mechanics that has prevented to date significant solutions to our decaying environment is its time reversibility, here referred to the invariance of the dynamical equations under time reversal technically expressed as the invariance of the Schroedinger's and Heisenberg's equations under anti-Hermiticity. Time reversal invariance is effective and necessary for the consistent representation of stable structures with point-like constituents, such as atomic structures. However, due to their known time irreversibility, the same invariance is grossly insufficient for the representation of any physical or chemical energy releasing process, such as nuclear fusions or fuel combustions. Moreover, recent studies in this book have shown that the EPR completion of quantum mechanics into a time-irreversible theory directly impacts the engineering realization of energy-releasing processes. In conclusion, a time-irreversible non-potential completion of quantum mechanics for extended constituents appears to be most promising for energy-releasing processes, while admitting as a particular case a time-reversible and non-potential completion for the representation of stable structures with extended constituents (Sect. 2.7, p. 51 of [1-3]).

1.5-5 One-side modular structure

From a mathematical viewpoint, the Schroedinger representation of quantum mechanics has a right modular structure in the sense of being characterized by the multiplication ordered to the right of a Hamiltonian

$H(r, p)$ to a state $|\psi(r)>$ of a Hilbert space, resulting in the familiar eigenvalue equations $H(r, p)\ |\psi(r)> = E\ |\psi(r)>$, while ignoring the corresponding left modular action $<\psi(r)|\ E = <\psi(r)|\ H(r, p)$ because trivially implied by the right action for stable systems of point-like constituents. As we shall see in Chapter 3, the relaxation of the above structure and the use of a full, right, and left bi-modular structure offers realistic possibilities for an axiomatic representation of time irreversibility for energy-producing processes with extended constituents (Sect. 3.4, p. 55 of [1-3]).

1.5-6. Single valuedness

Quantum mechanics has a single-valued structure in that all multiplications yield one result irrespective of the multiplication order. For instance, the product of the number 2 by 3 yields the single result 6 irrespective of whether 2 multiplies 3 to the right or 3 multiplies 2 to the left, and the same holds for all other quantum mechanical multiplications. In Santilli's view, the indicated single valuedness is sufficient for contemporary physical and chemical studies but grossly insufficient for the study of biological structures since the correlation/entanglement of two atoms in a DNA, mathematically representable with a product, may yield a complete organ with 10^{30} atoms. This occurrence suggests the need for biological structures of the most complex mathematics conceivable by the human mind, such as the time-irreversible hyperstructures whose product yields an unlimited number of ordered results (Sect. 3.5, p. 55 of [1-3]).

1.5-7. Positive energies

According to 20th century relativistic quantum mechanics, all particles identified to date, including antiparticles, have positive masses to prevent a violation of special relativity. However, as we shall see in Chapter 13, the demolition into the light of a pair of particles and its antiparticle both having positive energy is prohibited by the axioms of the theory, therefore supporting Paul Dirac's 1928 conception of antiparticles as having negative energies under the use of an appropriate mathematics expected to be based on a negative unit as a condition to prevent causality problems.

1.6 Santilli's rebuttal of Bohr's objections

A theory should be considered complete for given physical conditions when it provides an exact and time invariant representation of the physical reality characterized by measured experimental data. Quantum mechanics provides an exact and invariant representation of the experimental data for the hydrogen atom, accelerator particles, crystals, and other systems of point-like particles in a vacuum. Still, it cannot be considered a complete theory because it does not provide an exact and invariant representation of experimental data of nuclear physics 1.4-1 to 1.4-1-7, besides being disproved by widely ignored experimental evidence in virtually all fields of physics, such as * see the recent general review by E. Velardo [1-21]):

1.6-1) Three direct experimental verifications of the EPR argument that quantum mechanics is not a complete theory [1-6] [1-7] [1-8].

1.6-2) Experimental evidence [1-9] [1-10] on the lack of exact character of quantum electrodynamics for the muon data, and their sole known exact and invariant representation via hadronic mechanics [1-11] [1-12].

1.6-3) Experimental violation of quantum mechanics in the behavior of the mean life of the K-mesons with speed and its sole exact and invariant representation via hadronic mechanics [1-13]-[1-15].

1.6-4) Exact and invariant representation by hadronic mechanics of the shape and density of the proton-antiproton fireball in the Bose-Einstein correlation [1-16] [1-17] in lieu of the four completely arbitrary chaoticity parameters of quantum formulations.

1.6-5) Experimental confirmation in the USA and in Europe of the local character of the speed and frequency of light propagating within physical media [1-18], experimental confirmation on Earth of Zwicky's Tired Light [1-20] [1-21], with ensuing removal of Earth at the center of the universe which is inherent in the expansion of the universe due to the validity of Hubble's law on the cosmological redshift in all radial directions from Earth (www.eprdebates.org/no-universe-expansion.php).

References of Chapter 1

[1] Einstein, A., Podolsky, B. and Rosen, N.: Can quantum-mechanical description of physical reality be considered complete?, *Phys. Rev* 47, 777-70 (1935), http://www.eprdebates.org/docs/epr-argument.pdf

[2] Bohr, N.: Can Quantum-Mechanical Description of Physical Reality Be Considered Complete? *Physical Review* 48, 696-702 (1935), https://santilli-foundation.org/docs/santilli-532.pdf

[3] Santilli, R. M.: Lie-isotopic representation of stable nuclei I: Apparent insufficiencies of quantum mechanics in nuclear physics, *Ratio Mathematica,* 52, 43-63 (2024),

[4] Santilli, R. M.: *Overview of historical and recent verifications of the Einstein- Podolsky-Rosen argument and their applications to physics, chemistry, and biology,* APAV - Accademia Piceno Aprutina dei Velati, Pescara, Italy (2

[5] Beghella-Bartoli, S. and Santilli, R. M., Editors: *Proceedings of the 2020 Teleconference on the Einstein-Podolsky-Rosen argument that 'Quantum mechanics is not a compete theory,'* 941 pages, Curran Associates, New York, NY (2021)

[6] Fadel, M. et al.: Spatial entanglement patterns and Einstein-Podolsky-Rosen steering in Bose-Einstein condensates, *Science* 360, 409–415 (2018),

[7] Colciaghi, P. et al.: Einstein-Podolsky-Rosen Experiment with Two Bose-Einstein Condensates, P*hys. Rev. X* 13 (2023),

[8] Aspect, A. et al.: Experimental Realization of Einstein-Podolsky-Rosen-Bohm Gedankenexperiment: A New Violation of Bell's Inequalities, *Phys. Rev. Lett.* 49, 91-94 (1982),

[9] Miller, J. P. et al.: Muon (g2): experiment and theory, *Rep. Prog. Phys.* 70, 795–881 (2007), news.fnal.gov/2021/04/first-results-from-fermilabs-muon-g-2-experiment-strengthen-evidence-of-new-physics

[10] [1-10] Santilli, R. M.: Apparent Unsettled Value of the Recently Measured Muon Magnetic Moment, *Progress in Physics* 18, 15–18 (2022),

[11] Santilli, R. M.: Representation of the anomalous magnetic moment of the muons via the Einstein-Podolsky-Rosen completion of quantum into hadronic mechanics, *Progress in Physics* 17, 210–215 (2021),

[12] Santilli, R. M.: Representation of the anomalous magnetic moment of the muons via the novel Einstein-Podolsky-Rosen entanglement in: H. M. C. Garcia, J. J. C. Guzman, L. H. Kauffman and H. Makaruk, Editors, Scientific Legacy of Professor Zbigniew Oziewicz, World Scientific (2024),

[13] Cardone, F., Mignani, R. and Santilli, R. M.: On a possible energy-dependence of the K^0 lifetime, Part I *J. Phys. G: Part. Phys.* 18, L141-L144 (1992),

[14] Cardone, F., Mignani, R. and Santilli, R. M.: On a possible energy-dependence of the K^0 lifetime. Part II *J. Phys. G: Part. Phys.* 18, L61-L65 (1992),

[15] Arestov, Yu., Santilli, R. M. and Solovianov, V.: Experimental evidence on the isominkowskian character of the hadronic structure, *Foundationof Physics Letters* 11, 483 (8),

[16] Santilli R. M. Nonlocal formulation of the Bose-Einstein correlation within the context of hadronic mechanics, *Hadronic Journal* 5, 1–50 and 15, 81–133 (1992),

[17] Cardone F. and Mignani R. Nonlocal approach to the Bose-Einstein correlation, *JETP* 83, 435 (1996), www.santilli-foundation.org/docs/Santilli-130.pdf

[18] Ahmar H., Amato G., Kadeisvili J. V., Manuel J., West G. and Zogorodnia O. Additional experimental confirmations of Santilli's IsoRedShift and the consequential lack of expansion of the universe, of Computational Methods in Sciences and Engineering 13, 321– 375 (2013),
//www.santilli-foundation.org/docs/IRS-confirmations-212.pdf

[19] Santilli R. M. Experimental Verifications of IsoRedShift with Possible Absence of Universe Expansion, Big Bang, Dark Matter and Dark Energy, *The Open Astronomy Journal* 3, 124-132 (2010),
www.santilli-foundation.org/docs/Isoredshift-Letter.pdf

[20] Mignani, R. Quasars redshift in isominkowski space, *Physics Essay* 5, 531 (1992), http://www.santilli-foundation.org/docs/Santilli-31.pdf

[21] E. Velardo, An introduction to iso-, geno- and hyper- mathematics and their applications in physics, chemistry and biology, *Ratio Mathematica,* in press in Vol. 54 (2025),
www.santilli-foundation.org/docs/Velardo-2025

CHAPTER 2

OVERVIEW OF HADRONIC MECHANICS AND ITS APPLICATIONS

2.1. Introduction

Jeremy Dunning-Davies begins his paper [2-1] by quoting a profound statement from Paul Dirac, a Nobel Prize-winning physicist. Dirac suggested that the difficulties faced by quantum mechanics could only be resolved by a drastic change in the foundations of the theory, similar to the transition from Bohr's orbit theory to quantum mechanics. This sets the stage for the discussion on the limitations of quantum mechanics and the need for new approaches.

Karl Popper, [2-2], a prominent philosopher of science (see also books [2-3] and [2-4]), is known as a critic of the Copenhagen Interpretation of quantum mechanics. Popper argued that the observer's role in quantum mechanics should be the same as in classical physics, essentially to test the theory rather than influence it. Dunning-Davies highlights that Popper's attention to Santilli's work in his 1982 book [2-2] significantly endorsed Santilli's ideas. Popper noted that Santilli revived the first structure model of the neutron, which had been discarded due to the limitations of quantum mechanics. This revival was possible because Santilli applied generalized mathematics and mechanics rather than conventional quantum mechanics.

Popper's [2-2] supports Santilli's distinction between atomic and nuclear mechanics as providing a strong foundation for the EPR completion of quantum mechanics. Popper's view that quantum mechanics should not be universally accepted without new tests in nuclear and hadronic physics aligns with Santilli's work and calls for a reassessment of established theories, particularly given our known environmental problems.

2.2. Limitations of Quantum Mechanics

Dunning-Davies delves into the limitations of quantum mechanics, emphasizing that Santilli devoted his life to extending the theory to cover situations where it was not applicable. He argues that many of these limitations are hidden from the public and students, placing Santilli in opposition to the established scientific community. This opposition is compared to historical figures like Waterston, whose work on the kinetic theory of gases was ignored, delaying scientific progress for a century.

Santilli's work is described as challenging the bases of quantum mechanics and relativity, asserting their incompleteness. This challenge has disadvantaged him in the scientific world, where questioning fundamental theories is still perilous. Dunning-Davies underscores the importance of Santilli's investigations, which seek to address the gaps in conventional theories.

2.3. The Need for New Clean Nuclear Power

The discussion shifts to the pressing issue of clean energy, highlighting Santilli's investigations into new clean energies. Dunning-Davies points out that the depletion of energy stocks and environmental concerns make this research particularly urgent. He notes that the future energy demand, especially from developing regions like the Indian subcontinent and Africa, will escalate enormously, making nuclear power a necessary option despite its controversial nature.

Dunning-Davies argues that alternatives like solar, wind, geothermal, and wave energy are insufficient to meet future energy demands. He asserts that nuclear power is the only realistic solution with its potential for high energy output. However, safety concerns surrounding nuclear power plants and nuclear waste disposal remain significant challenges.

2.4. The Need for Recycling Nuclear Waste

Santilli's work on the safe in-house recycling of nuclear waste via their transmutation into short life isotopes is presented as a groundbreaking solution. Dunning-Davies explains that Santilli's theoretical method involves exposing radioactive waste to various low energy radiations, including a coherent flow of photons with the required resonating frequency, achievable with a synchrotron, to stimulate nuclear neutrons to decay. This method, if validated, could revolutionize nuclear waste management.

The reluctance to perform the necessary experiments to validate Santilli's theory despite their limited cost is attributed to its conflict with conventional wisdom and the vested interests of those profiting from current waste disposal methods. Dunning-Davies argues that the potential benefits of successful implementation far outweigh the costs and that Santilli's theory has already led to the production of the new clean Hydrogen fuel magnegas synthesized from petroleum, the industrial synthesis of the neutron from the Hydrogen and other technologies demonstrating its practical potential.

2.5. Rutherford's Synthesis of the Neutron in a Star

Dunning-Davies revisits Rutherford's early 20th century proposal of a neutron as a compressed hydrogen atom and the subsequent experimental confirmation by Chadwick. Despite being initially dismissed due to inconsistencies with quantum mechanics, Santilli revived this model using hadronic mechanics, resolving the issues that led to its abandonment.

Santilli's model of the neutron as a generalized bound state of an electron totally compressed inside the proton is highlighted as a significant achievement, providing a consistent explanation for various neutron characteristics that quantum mechanics could not. This success is attributed to the generalization of Heisenberg's uncertainty principle within hadronic mechanics.

Dunning-Davies emphasizes that this model explains all neutron's properties, including spin, electrical neutrality, anomalous magnetic moment, rest energy, mean life, charge radius of about 1fm, and spontaneous decays. Santilli's work extends this model to all leptons, mesons, and baryons, providing a comprehensive new framework for understanding particle physics.

2.6. The Need to Generalize Heisenberg's Uncertainty Principle

Central to Santilli's work is the generalization of Heisenberg's uncertainty principle along the EPR argument he achieves for extended particles or wave packets in conditions of deep mutual penetration. Dunning-Davies explains that this generalization is crucial for the new method of recycling nuclear waste and for much-needed advances in nuclear fusions. Santilli's generalization of Heisenberg's uncertainty principle has enormous implications for physics, potentially overturning conventional views on the structure of physical reality and opening new avenues for clean energy research.

Figure 2-1. In view of the known prohibition to remove radioactive nuclear waste from power plants, their sole possible recycling is their transmutations by the nuclear power plants themselves into isotopes with short mean lives. The biggest environmental damage caused by the wide academic acclaim of the terminal character of quantum mechanics for all of the universe, is that of prohibiting research in the indicated recycling via the discredit of the proponents despite its full admission by the EPR completion of quantum into hadronic mechanics.

2.7. Recycling Nuclear Waste via Stimulated Decay

Dunning-Davies details Santilli's proposal for safely disposing of nuclear waste through stimulated decay, using an intense photon flow to induce rapid transmutation of unstable nuclei. He presents the example of transmuting uranium to plutonium with a significantly shorter half-life, highlighting the potential for drastically reducing the long-term storage requirements of nuclear waste.

The feasibility of this method depends on experimental validation, which Dunning-Davies argues should be prioritized given the potential ecological benefits for our rapidly decaying environment. Despite opposition from established interests, he calls for the scientific community and national governments to fund the necessary experiments to test Santilli's thesis.

2.8. The Einstein, Podolsky, and Rosen argument

Dunning-Davies discusses the famous 1935 paper by Einstein, Podolsky, and Rosen, which questioned the completeness of quantum mechanics. He reviews the extensive experimental and theoretical efforts to address the argument and introduces Santilli's 1998 resolution using his iso-mathematics framework.

Santilli's hadronic mechanics is an axiom-preserving completion of the local, linear, and potential structure of quantum mechanics for point-like particles for the additional representation of non-local, non-linear, and non-potential interactions for extended nucleons in the condition of mutual penetration as occurring in a nuclear structure. (Fig. 1.3) Under the latter conditions, Heisenberg's uncertainty principle is generalized into a form progressively recovering Einstein's determinism which permits exact and invariant representations of nuclear data. Dunning-Davies emphasizes the need for the scientific community to examine Santilli's proposals without prejudice, given their potential to resolve longstanding issues in quantum mechanics

Figure 2-2. In Santilli's view, the biggest insufficiency of quantum mechanics has been the century-old inability to achieve a representation of the synthesis of the neutron from the Hydrogen in the core of stars because prohibited by Heisenberg's uncertainty principle as well as from the fact that the rest energy of the neutron is 0.782 MeV "bigger" than the sum of the masses of the proton and the electron, thus implying a "mass excess" (rather than the familiar "mass defect") beyond any hope of quantum mechanical treatment.

2.9. The Limitations of von Neumann's, Bell's, and Other Theorems

Dunning-Davies explains how Santilli's new mathematics renders von Neumann's theorem on hidden variables inapplicable and challenges the universality of Bell's inequality, which holds specifically for conventional quantum mechanics. He highlights recent experimental support for the EPR argument and discusses the implications of these findings for the future of physics.

Dunning-Davies notes that thought experiments, such as the one leading to the EPR paradox, should be viewed cautiously due to their inherent assumptions. He describes the recent experimental work by the Basel team, which used lasers to cool atoms and measure spin correlations, providing empirical support for the EPR argument.

Dunning-Davies advocates for independent experimental validation of Santilli's technologies, highlighting their potential to revolutionize our understanding of nuclear physics and energy. He underscores the importance of open-mindedness in scientific inquiry and the need to re-evaluate established theories in light of new evidence.

A general review of the mathematics underlying hadronic mechanics and its applications in physics, chemistry and biology is available in Ref. [2-5].

References of Chapter 2

[1] Dunning Davies, J.: Foreword for isotopic generalizations of quantum theories, Ratio Mathematica, **52,** 9-20 (2024), https://www.santilli-foundation.org/docs/1-RM-SI-Vol-52-2024-fin.pdf

[2] Popper, K.: *Quantum Theory and the Schism in Physics,* Cambridge University Press (1982).

[3] Dunning Davies, J.: *Exploding a myth: "Conventional Wisdom" or Scientific Truth,* Horwood Publishing Limited, London, UK (2007).

[4] Horgan, J.: *The End of Science*, Basic Books, New York, NY (2015).

[5] Velardo, E.: An introduction to iso-, geno- and hyper- mathematics and their applications in physics, chemistry and biology, *Ratio Mathematica,* in press (2025), https://www.santilli-foundation.org/docs/Velardo-2025.pdf

[6] Santilli, R. M.: An introduction to new sciences for a new era, \textit{Clifford analysis, Clifford algebras and their applications} {\bf 6(1)}, 1-119 (2017),
\\http://www.santilli-foundation.org/docs/new-sciences-new-era.pdf

CHAPTER 3

THE INCEPTION OF HADRONIC MECHANICS

3.1. Introduction

In The Italian-American scientist R. M. Santilli has dedicated his research life to the mathematical, theoretical, experimental and industrial verification of Einstein-Podolsky-Rosen argument that *"quantum mechanics is not a complete theory"* [1-1] (see also his recent talk in Cambridge, UK, delivered on April 28, 2025, www.santilli-foundation.org/docs/Cambridge-Lecture-2025.mov).

The research initiated with Santilli's Ph. D. studies at the University of Torino, Italy, on grounds that quantum mechanics is a time reversal invariant theory, thus being unable for a consistent representation of time irreversible processes in physics, chemistry and biology (Fig. 3-1). This insufficiency can be seen from the invariance of Heisenberg's time evolution of an observable A under anti-Hermiticity, i dA/dt = [A. H] = A H - H A = - $[H, A]^\dagger$ with ensuing conservation of the energy, i dH/dt = 0.

To overcome the above insufficiency, Santilli proposed in his Ph. D. Thesis [3-1] [3-2] [3-3] the generalization of Lie algebras into broader algebras called Lie-admissible with brackets (A, H) = A < H – H > A = A R H – H S A where the R and S operators are nowadays called the forward and backward Santillians [3-26]-[3-29], which verify the following properties:

1) The new brackets admit the Lie brackets as a particular case as a necessary condition for the new mechanics to admit quantum mechanics as a particular case.

2) The new brackets are such that their antisymmetric form (A, B) - (B, A) verify the Lie axioms as a necessary condition to be a completion of the Lie brackets.

3) The new brackets characterize the broader Heisenberg-Santilli Lie-admissible time evolution i dA/dt = (A, H) which is no longer invariant

under time reversal (for $R \neq S$) and characterizes the time rate of variation of the energy, idH/dt = (H H)≠ 0 as a necessary condition to represent irreversible processes.

Following a one-year visiting position at the Center for Theoretical physics of the University of Miami, Florida, under NASA support, the position of Associate Professor of physics at Boston University under research support by the U. S. Air Force, and a visiting position at the Center for Theoretical Physics of MIT, -Santilli joined the faculty of the Lyman Laboratory of Physics of Harvard University on September 8, 1977. On the day of his arrival, Howard Georgi, his supervisor at Harvard University, received a phone call from David Peaslee of the U. S. Department of energy inviting Harvard University to apply for the administration of a grant in support of Santilli's studies on the irreversibility of nuclear fusions. This invitation, done under a great American President, the nuclear physicist Jimmy Carter, was motivated by Santilli's 1965 Ph. D. thesis in the field [3-1] [3-2] [3-3].

The DOE proposal was accepted with appreciation, although Santilli pointed out to Peaslee that he could not study energy-releasing processes without first achieving a quantitative representation of the most fundamental synthesis in nature, that of the neutron from a proton and an electron in the core of stars. In turn, such a representation required the development of new mathematics to represent particles' dimensions, shape density, and non-potential interactions caused by deep mutual entanglement.

Peaslee accepted Santilli's proposal, and his affiliation was then finalized at the Department of Mathematics of Harvard University as an associate of Shlomo Stenberg. This interim initiated DOE contracts ER-78-S-02-47420.A000, AS02-78ER04742, DE-ACO2-80ER10651, DE-ACO2-80ER-10651.A001, and DE-ACO2-90ER10651.A002 that were administered by Harvard University until August 1981 and then administered by the Institute for Basic Research, then located in a Victorian on Harvard's Grounds.

In this chapter, we outline the most important aspects of the voluminous literature that originated from the indicated DOE grants and published in the period 1978-1983 on the conception of hadronic mechanics as a Lie-admissible completion of quantum mechanics for the representation of irreversible physical, chemical and biological processes (Fig. 3.1), which literature comprises:

i) The initial 202-page memoir [3-4] and 328-page memoir [4-5] for the classical and operator foundations, respectively, of the Lie-admissible and Lie-isotopic mathematics and mechanics with the first known application to the representation of all characteristics of mesons and leptons as generalized hadronic bound states with physical constituents.

Figure 3-1. In support of the EPR argument [1-1], Santilli studied nuclear fusions, chemical combustion and an embryo that cannot be consistently represented via quantum mechanics due to its strict reversibility over time compared to the known irreversibility of the systems considered and constructed an irreversible completion of quantum mechanics under the proposed name of hadronic mechanics [3-1]-[3-7].

ii) Monographs [3-6] [3-7] on the foundations of the Lie-admissible mathematics, also called geno-mathematics, including the Lie-admissible generalization of Lie's theory, mechanics and geometries.

iii) Monographs [3-8] [3-9] on the generalized, time-reversible Lie algebras attached to the Lie-admissible algebras called Lie-isotopic algebra or iso-algebras and related Lie-isotopic mechanics or iso-mechanics used for the exact and invariant representation of the experimental data of stable, thus reversible nuclei.

iv) Workshops on Lie-Admissible Formulations [3-10] [3-11]; First International Conference on Nonpotential Interactions and their Lie-Admissible Treatment [3-12]; Workshops on Hadronic Mechanics [3-13] [3-14]; Preprint series Applications of Lie-admissible algebras in physics [3-15]; reprint volume on Irreversibility and Non-potentiality in Statistical Mechanics, [3-16]; Tomber's Bibliography and Index on Nonassociative Algebras [3-17].

v) Representative papers [3-18]-[3-25].

3.2. Experimental foundations

According to clear experimental measurements, atomic structures are composed of point-like electrons at large mutual distances from the nuclei. By comparison, nuclear structures comprise extended and dense charge distributions of protons and neutrons (collectively called nucleons) in conditions of partial mutual penetration. For example, Helium-4 has the charge radius R = 1.67 fm, which is 0.07 fm smaller than the proton's diameter, and the neutron D=1.74 fm. Consequently, the four nucleons composing the Helium-4 are in conditions of 0.07 fm mutual penetration of their dense charge distributions plus deep mutual entanglement of their wave packets by the EPR argument [1-1] (Fig. 1-3). For heavier nuclei, the above conditions are magnified by the strength of nuclear forces, as a result of which the wave packets of protons and neutrons are in conditions of essentially total mutual penetration/entanglement.

3.3. The new non-Hamiltonian interactions

The above experimental data suggest that strong nuclear interactions are characterized by:

Conventional action at a distance, local, linear, and potential interactions that, as such, are fully representable by a conventional Hamiltonian H, plus

Contact, thus zero-range interactions that are non-local (in the sense of being distributed over a volume) as pioneered by L. de Broglie and D. Bohm; non-linear (in the sense of depending on powers of the wave functions) as pioneered by W. Heisenberg; and non-potential (in the sense of not being derivable from a potential) as pioneered by R. M. Santilli [3-4]-[3-8]. As such, the latter interactions cannot be represented with a Hamiltonian.

The research under the indicated DOE grants initiated with the systematic studies of potential and non-potential forces in Newton's equations via *variational selfadjointness*, providing the necessary and sufficient conditions for given forces to be derivable from a potential. Forces verifying these conditions are called variationally selfadjoint and indicated with the letters SA, while forces violating said conditions are called variationally non-selfadjoint and indicated with the letters NSA. Santilli first presented these studies in a Seminar Course delivered at the Lyman Laboratory of Physics in the Spring of 1978, first published in Sect. 2, p. 239 of memoir

[3-4], and then systematically studied in the Springer-Verlag monograph [3-8].

To understand the significance of the studies, we should note that Santilli's non-potential interactions can be a realization of the missing fifth interactions.

3.4. Representation of non-Hamiltonian interactions

3.4.1. Classical representation via Birkhoffian mechanics.

Santilli's research at Harvard University under DOE support began in the late 1970s with the first known representation of the most general possible (well behaved) non-linear and non-local Newton's equations of motion $m(dv/dt)-F^{SA}(r)-F^{NSA}(t,r,p,...) = 0$ via analytic equations derivable from a variational principle because necessary for operator images [3-8] [3-9] .

By using the unified coordinates in phase space $a = (a^{\mu}) = (r,p) = (r^k,p_k)$, $\mu = 1, 2, 3, ...6$, $k = 1, 2, 3$), the Hamiltonian $H(t, r, p) = H(t, a)$, and the partial derivatives $\partial_{\alpha} = \partial/\partial a^{\alpha}$, $a^{\alpha} = t, a^{\mu}$, the general Newton's equations can be represented via the true Hamilton's equations (those with external terms) $\omega_{\mu\nu}(da^{\nu}/dt) - \partial_{\mu}H(t,a) - F^{NSA}(t,a,...) = 0$, where ω is the canonical tensor with off diagonal terms -1, $+1$ (p. 15 of [3-9]). However, the true Hamilton's equations are NSA and, therefore, they cannot be derived from a variational principle. Therefore, they cannot be used as the classical analytic foundations of hadronic mechanics.

In view of the above limitations, Santilli selected Birkhoff's equations $\{[\partial_{\mu}R(t, a)_{\nu} - \partial_{\nu} R(t, a)_{\mu}]\, da^{\nu} /dt - (\partial_{\mu}B(t, a) + \partial_t R(t, a)_{\mu}\}^{SA} = 0$ (Eqs. (4.2.1), p. 30 of [3-9]) which are SA, thus being derivable from a variational principle with action $A = \int dt[R_{\mu}(t, a)\, (da^{\mu}/dt) - B(t, a)]$ (Eqs. 4.2.15, p. 32 of [3-9]), where B(t,a) is the Birkhoffian, namely a non-linear and non-local generalization of the Hamiltonian that no longer represents the total energy, and the functions $R_{\mu}(t, a)$ are restricted by the condition $\det.(\Omega_{\mu\nu}) = \det.\{[(\partial_{\mu}R(t, a)_{\nu} - \partial_{\nu} R(t, a)_{\mu})]\} \neq 0$. It is evident that the truncated Hamilton's equations (i.e., the S A equations without external terms) are a particular case of Birkhoff's equations for $R_{\mu}(t, a) = 0$, $\mu = 1, 2, 3$ and $R_{\mu}(t, a) = p_k$, $\mu = 3, 4, 5$. Birkhoffian mechanics is the generalization of all functional, algebraic and geometric aspects of Hamiltonian mechanics for Birkhoff's equations [3-9].

The main significance of Birkhoff's equations is that of allowing the first known analytic representation of all possible contact, non-linear, non-local and non-potential interactions of extended bodies moving within physical media (Sect. 4.5, p. 54, on) "Direct Universality of Birkhoff's Equations" [3-9]), with significant applications in various fields.

As an illustration in physics, Birkhoff's equations allow the only known classical representation of particle/wave packet entanglements with important applications to the measurement theory, Einstein's determinism [1-1] and other fields, while in engineering, the action principle of Birkhoff's equations and related optimal control theory allows the only known optimization of the actual shape of a wing under velocity-dependent resistive forces compared to the current optimization via a set of isolated points.

More generally, Birkhoffian mechanics is recommended for the representation of all motions of extended bodies within physical media since they all depend on a quantitative representation of contact NSA forces which is done by the difference between the Birkhoffian and the canonical tensors, $\Omega_{\mu\nu} - \omega_{\mu\nu}$.

On algebraic grounds, the Hamiltonian time evolution of an observable A in terms of t he Poisson-Lie brackets $(dA/dt) = \partial\mu A\ \omega\mu\nu\ \partial\nu H = [A,H]$ is generalized into the form $(dA/dt) = \partial\mu A\ \Omega\mu\nu\ \partial\nu H = [A,H]*$ whose new brackets $[A,H]*$ verify all Lie axioms, yet they are structurally more general than the Poisson brackets. Hence, they characterized the isotopies of the various branches of Lie's theory, nowadays known as the Lie-Santilli isotheory (first presented in Charts 5.1 to 5.5 of [3-9]), which is at the foundation of the generalization of the Galileo and special relativities for extended particles moving within physical media (Chapter 7).

On geometric grounds, Birkhoffian mechanics has triggered the isotopies of the geometries associated to Hamiltonian mechanics, such as the iso-Euclidean iso-symplectic and other broader geometries [8-6] [8-7]. For the operator map of Birkhoffian mechanics into hadronic mechanics.

3.4.2. Operator representation via the enveloping algebra

Santilli initiated his search for an operator representation of non-Hamiltonian interactions by generalizing the universal enveloping algebra U of quantum mechanics into a non-associative algebra U* with a Lie-admissible product (A, H) = A < H – H > A = ARH – HSA [3-4 [3-5]. Following numerous attempts via the use of all known non-associative algebras, Santilli had to abandon this line of research because of mathematical and physical inconsistencies similar to those that have prevented the use in physics of octonions due to their non-associative character (Chapter5).

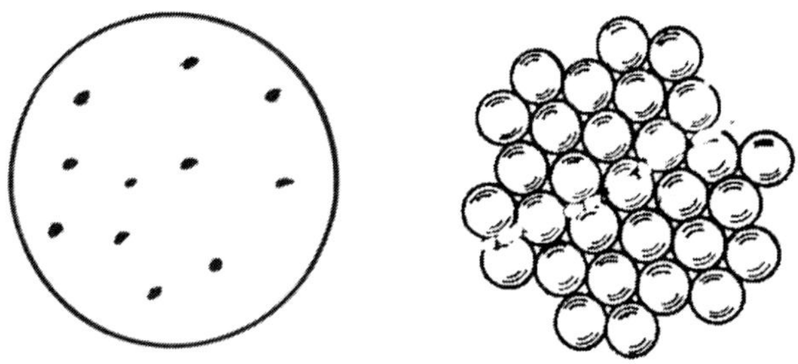

Figure 3-2. In view of its strict locality, quantum mechanics can essentially represent nuclei as ideal spheres with point-particles in its interior under sole action-at-a-distance potential interactions (left view). Hadronic mechanics has been built to achieve a time invariant representation of nuclei as they are in the physical reality, i.e. as a collection of extended and hyperdense nucleons in conditions of partial mutual penetration (Fig.1-3) with ensuing potential/Hamiltonian as well as contact/non-Hamiltonian interactions (right view).

In this way, Santilli was left with no other option than searching for an axiom-preserving completion of the basic axioms of quantum mechanics since, in that case, the absence of catastrophic inconsistencies was assured. Following due research, Santilli discovered the completion of the associative product AB into the associativity preserving form A∗B = ASB, (A∗B)∗C = A∗(B∗C), where the quantity S is a new invertible function or operator. The new product A∗B was first presented at the classical level in Eq. (3.6.11), p. 333 of [3-4], then at the operator level in Eq. (4.14.23a), p. 723 of [3-5], and studied systematically in Chart 5.2, p. 154 of [3-9]. Note that the completion of the product AB → A∗B implies the completion of the unit 1 → I *= 1/S with corresponding identity law I^*∗A = A∗I = A.

The new quantity S was initially called the *isotopic element* and, forty-six years later, it is called the *Santillian* [3-26]-[3-29]- with explicit realizations of the type S = Diag. (a, b, c, d)Exp{-f(t, ψ, ...)} << 1, f>0, where: a, b, c represent the semi-axes of the particle considered; d represents its density; ψ represents the wave function of the corresponding quantum systems; and the function f(t, ψ, ...)} represents all non-linear, non-local and non-potential interactions. Consequently, as it is the case for the Hamiltonian, H the Santillian S is fixed for a given system.

Once assured by the axiomatic consistency due to the preservation of the associative character of the envelope, Santilli introduced the name *hadronic mechanics* in the abstract and page 614 of the memoir [3-5] to denote its primary application to strongly interacting particles called *hadrons* while smoothly and unequivocally recovering quantum mechanics for sufficiently big mutual distances represented by then limit S = 1.

Soon following the discovery of the new associative product, Santilli realized that the completions AB → A∗B = ASB and 1 > I *= 1/S imply a compatible completion of the entire 20th-century applied mathematics, including numbers, functions, spaces, differential, calculus, algebras, geometries, symmetries, etc., with no exception known to this author. At this point, we merely indicate the need of avoiding the impression of inconsistencies in hadronic mechanics, which occur when the properties of the new product A∗B are elaborated with the product AB.

Additionally, the completion AB → ASB and 1→ I^* = 1/S generally produces different new mathematics depending on the topological characteristics of S. The studies under the DOE grants indicated in Sect. 3.1 were conducted via the use of the following new mathematics and related branches of hadronic mechanics:

1) *Lie-admissible* or *geno-mathematics* for the invariant representation of *irreversible* processes between extended hadrons with Hamiltonian and non-Hamiltonian interactions, such as nuclear fusions of combustion. The corresponding new mechanics is called *Lie-admissible or geno*-mechanics [3-1]-[3-9] and Fig. 3.2.

2) *Lie-isotopic or iso-mathematics* for the invariant representation of *closed reversible* systems of extended hadrons with Hamiltonian and non-

Hamiltonian internal forces, such as nuclear structures or complex molecular clusters. The corresponding new mechanics is called *Lie-isotopic or iso-mechanics [3-4]-[3-9].*

The classification of all new mathematics and corresponding branches of hadronic mechanics, which can be derived from completions $B \rightarrow A*B = ASB$ and $1 \rightarrow I^* = 1/S$ is outlined in Chapter 4.

3.5. The inception of hadronic mechanics

3.5.1. Introduction

Recall that the primary aim of the studies under the indicated DOE grants was the identification of a consistent and time-invariant representation of nuclear fusions between nuclei composed of extended protons and neutrons in conditions of deep mutual entanglement with resulting conventional potential as well as contact non-potential internal forces. An evident pre-requisite for the above aim was the time-invariant representation of isolated stable nuclei with the indicated extended constituents and internal forces.

In the 1960's, a primary objection against the indicated representation was and remains to a considerable extent, that composite systems with non-potential internal forces do not admit a physically consistent operator form because non-conservative forces do not conserve the total energy. Santilli disproved this view via the conditions of variational selfadjointness and the proof that under certain conditions, non-conservative internal forces do allow the conservation of total quantities (see Eqs. (6.3.36), p. 236 [3-9]), by therefore establishing that the indicated conception of stable nuclei does indeed admit a fully Newtonian counterpart. The understanding is that the rigorous proof of total conservation laws requires the knowledge of symmetries under non-potential interactions, which requires a step-by-step corresponding generalization of Lie's theory (see Sect. 6.3, p. 225 of [3-9]).

In this section, we guide the general physics audience on the inception of hadronic mechanics, first at the classical level and then at its operator counterpart.

3.5.2. Operator image of Birkhoffian mechanics

As reviewed in Sect. 2.3.1, the Birkhoffian necessarily depends on both coordinates r and moments p, $B = B(t,a) = B(t,r,p)$ [3-9]. Consequently, its operator image obtained via the naive use of the Birkhoffian variational principle is given by $i\partial_t|\psi(r, p)\rangle = B(t, r, p)|\psi(r, p)\rangle$, namely, the "wave functions" of the operator image of Birkhoffian mechanics must depend on both coordinates r and momenta p.

Unfortunately, possible Hilbert spaces with states $|\psi(r, p)\rangle$ depending on both coordinates r and moments p are afflicted by so many consistency problems to prevent the EPR completion of quantum mechanics, such as [3-9]: loss of the Fourier transform of the wave function from the ordinate space $|\psi(r)\rangle$ to the momentum space $|\psi(p)\rangle$; loss of the canonical commutation rule $[r, p] = ih^-$ in favor of its commutativity $[r, p] = 0$; consequential loss of the Lorentz and other space time symmetries; lack of true operator image due to the difference between classical and operator commutation rules; and other inconsistencies.

To resolve these inconsistencies, Santilli pointed out in Eq. (14), p. 259 of [3-9] that the Birkhoffian is the product of the Hamiltonian and the Santillian, B = H S with consequential embedding of all non-linear, non-local and non-potential interactions in the core invariant of hadronic mechanics, the isounit $1^* = 1/S$, resulting in the *consistent operator image of Birkhoff's equations* $B(t, a)|\psi(r)\rangle = H(r, p) \star |\psi(r)\rangle = H(r, p)\ S(t, r, p, \psi, ...)\ |\psi(r)\rangle = E\ |\psi(r)\rangle$ which is at the foundation of hadronic mechanics (Chart 6.1, p. 253 of [3-9]).

3.5.3. Inception of Santilli's Lie-Admissible Branch of Hadronic Mechanics

To honor Einstein's [1-1] completion of quantum mechanics, Santilli wanted hadronic mechanics to be characterized by *new realizations of the same abstract axioms of quantum mechanics* to such an extent that the two mechanics had to coincide at the abstract, realization-free level and their differences had to appear solely in specific realizations. This condition would also allow hadronic mechanics to recover quantum mechanics smoothly, uniquely, and unambiguously at mutual distances bigger than the range of strong interactions.

Santilli achieved the above condition for the completion of the enveloping algebra U of quantum mechanics with product AB into the algebra U* with axiom-preserving product A*B =- ASB. However, achieving the same abstract identity for quantum and hadronic dynamical equations was much more difficult because it essentially required representing time-irreversible processes via the abstract axioms of time-reversible events.

Following several trials and errors, Santilli achieved the intended result in the 1979 paper [3-15] from the Department of Mathematics of Harvard University in which the universal enveloping iso-associative algebra U^* was divided into two inequivalent enveloping algebras, the first algebra $U^>$ characterized by all products ordered to the right $A>B = A\rightarrow S\rightarrow B$ and the second algebra $^<U$ characterized by all products ordered to the left $A<B = A\leftarrow R\leftarrow B$. Time irreversible processes in our physical reality, i.e., for evolution forward in time, are represented with all isoproducts ordered to the right $A>B = A\rightarrow S->B$ and relative forward time-dependent unit $I^>(t, \psi, \ldots) = 1/S^>(t, \psi, \ldots)$ while their time reversal image is represented with all products ordered to the left $A<B = A\leftarrow R\leftarrow B$ and relative backward time unit $^<I^>(- t, \psi, \ldots) = 1/R^>(- t, \psi, \ldots)$. A mathematically and physically rigorous axiomatization of irreversibility is then assured for $R\neq S$.

The corresponding axiomatization of dynamical equations is a mere consequence of the above-ordered products. In fact, the universal enveloping algebra of hadronic mechanics is given by $^<U\times U^>$ and implies the mapping of the two equivalent, thus reversible, Heisenberg's equations for Hermitean Hamiltonians $<\psi(- t, r)|\ E = <\psi(- t, r)|\ H(r, p)$ and $H(r, p)>|\psi(t, r)> = E\ |\psi(r, t)>$ into the *geno-Schroedinger equations,* also called *Schroedinger-Santilli geno-equations* (first introduced in Eq. (14), p. 299 of [3-9]) $<\psi(- t, r)|^<E = <\psi(-t, r)|<H(r, p) = <\psi(-t, r)|R(- t, \psi^*/\psi, \ldots)\ H(r, p)$ and $H(r, p)>|\psi(t, r)> = H(r, p)\ S(t, \psi^*/\psi, \ldots)\ |\psi(t, r)> = E^>\ |\psi(t, r)>$ on genostates $<\psi(-t, r)|$ and $|\psi(t, r)>$ of the geno-Hilbert space [3-23], which equations are clearly inequivalent since $^<E \neq E^>$ under the Hermiticity/observability of the non-conserved Hamiltonian, while preserving the abstract associativity of the envelope of quantum mechanics, separately, for motions forward and backward in time, thanks to the ordering of all products to the right and to the left, respectively.

Regarding the completion of Heisenberg's equation, the universal geno-associative algebra of hadronic mechanics $^<U\times U^>$ implies the following evolution of an observable in forward time $Q(t) = \exp\{HRti\}\ Q(0)\ \exp\{-itSH\}$ whose version for very small variations of time is given by *Santilli's Lie-admissible and Jordan-admissible generalization of Heisenebrg's equations*, also called Santilli geno-equations $idQ/dt =(Q, H) = Q<H - H>Q = QRH - HSQ$ (first introduced in Eq. (4.4.11), p. 719 of [3-5]) which can be decomposed in the form $idQ/dt = (Q, H) = (QTH - HTQ) + (QJH + HJQ) = [Q, H]^* + \{Q, H\}^*$, where $T = R + S > 0$, $J = - R + S > 0$. Consequently, brackets (Q, H) are jointly Lie-admissible because the brackets $[Q, H]^*$ verify the Lie-axioms, and Jordan-admissible because the brackets $\{Q, H\}^*$

verify the Jordan axioms [1-3]. Intriguingly, for R =- 1 and S = 1 - F/H, Santilli's geno-equations become idQ/dt = (Q, H) = Q<H – H>Q = QH - HQ + FQ, where F represents Lagrange's and Hamilton's external terms. The non-conservation of the Hermitian Hamiltonian is given by idH/dt = H<H – H>H =H(R-S)H $\neq$ 0. In particular, for R = 1 and S = 1 - F/H, idH/dt = H<H – H>H = H(R+S)H = HF, where F represents Lagrange's and Hamilton's external terms, thus confirming that the Lie-admissible dynamical equations of hadronic mechanics provide the missing operator version of Lagrange's and Hamilton's external terms F. Consequently, Lie-admissible formulations verify all three conditions for characterizing the needed completion of quantum mechanics indicated in Sect. 3.1. In particular, external terms are algebraically represented via the addition of Jordan brackets to Lie brackets.

In Sect. 6.3, p. 225 of [3-89], Santilli also provided a simple method for mapping any given reversible quantum model into its irreversible Lie-admissible counterpart. The method is given by the identification of two non-unitary transformations X, Y such that $XY^{\dagger}$= $I^{>}$= 1/S, $YX^{\dagger}$= $^{<}I$= 1/R, in which case 1 $\rightarrow$ X 1 $Y^{\dagger}$ = $I^{>}$ and AB $\rightarrow$ X(AB) $Y^{\dagger}$ = A' S B', S = $(X\, Y^{\dagger})^{-1}$, A' = X A $Y^{\dagger}$, B' = X B $Y^{\dagger}$. It should be stressed that, to avoid the catastrophic inconsistencies of quantum mechanical generalized theories, the resulting irreversible formulation has to be reformulated in terms of geno-unitary transformations, Eqs. (11), p. 259 of [3-9].

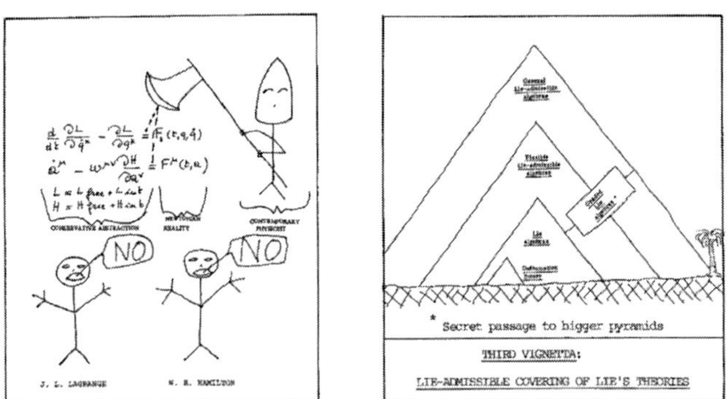

Figure 3-3. Contact non-potential forces experienced by extended bodies moving within a medium were first introduced by Newton with velocity dependent resistive forces, and represented by Lagrange and Hamilton with the external terms in their analytic equations. In this figure, we show two "vignettes" presented by Santilli at a meeting in May 1978 of the faculty of the Lyman Laboratory of Physics of Harvard University: the first (on the left) suggesting the completion of quantum mechanics

with the representation of Lagrange's and Hamilton's external terms via Lie-admissible algebras, and the second (on the right) illustrating the universality of such a representation since said algebras include as particular cases all algebras known in mathematics.

Intriguingly, by using the preceding decomposition of the Lie-admissible brackets for $R = 1$ and $S = 1 - F/H$, the Lie-admissible time evolution of the energy H acquires the form $i dH/dt = (H, H) = H<H - H>H = (HTH - HTH) + (HJH + HJH) = HJH + HJH = HF$, showing that *the operator form of Lagrange's and Hamilton's external terms can be represented by the addition of Jordan's brackets to the conventional Lie brackets.*

3.5.4. Inception of Santilli's Lie-Isotopic Branch of Hadronic Mechanics

In the early 1980s, a generalization of the classical Hamilton's equations for representing closed isolated systems with potential and non-potential internal forces did not exist. Therefore, it had to be constructed.

Santilli solved this additional problem by noting that the *iso-Schroedinger equations of hadronic mechanics,* also called *Schroedinger-Santilli iso-equations*, are the particular case $R = S$ of the Lie-admissible geno-equations given by $<\psi(r)|*H = <\psi| S H(r, p) = <\psi|E$, and $H(r, p)*|\psi> = H(r, p)S |\psi> = E'|\psi>$, $E = E'$ (first introduced in Eq. (14), p. 259 of [3-9]). Consequently, we have the following relationship between the Birkhoffian and the Hamiltonian $B|\psi> = HS|\psi> = E|\psi>$, namely, *the Birkhoffian is the product of the Hamiltonian and the Santillian*, $B = HS$ (Sects. 3.4.1 and 3.5.2)

The second fundamental equation of the Lie-isotopic branch of hadronic mechanics was easily derived from the assumption of the universal iso-associative enveloping algebra U^*, resulting in *Santilli's Lie-isotopic equation,* or Santilli's isoequation for short, $dQ/dt = [Q, H]^* = Q*H - H*Q = QSH - HSQ$, first formulated in Eq. (18a), p. 153, Ref. [3-9], where the Hamiltonian H represents all local, linear and potential interactions while the Santillian S represents all non-local, non-linear and non-potential interactions with the indicated realizations $S = Diag.(a, b, c, d) Exp\{-f(t, \psi^*/\psi, \ldots)\} << 1$, $f>0$.

The simple method indicated earlier for the mapping of any given reversible quantum model into its irreversible Lie-admissible counterpart acquires the simple form for the mapping of reversible quantum systems into the hadronic representation of the dimension, shape, and density of hadrons and their non-Hamiltonian interactions consisting in the identification of one non-unitary transformation X such $X.X^{\dagger} = \hat{I}^{>} = 1/S$, in which case $1 \rightarrow X\,1\,X^{\dagger} = \hat{I}^{>}$ and $AB \rightarrow X(AB)\,X^{\dagger} = A'\,S\,B'$, $S = (X\,X^{\dagger})^{-1}$, $A' = X\,A\,X^{\dagger}, B' = X\,B\,X^{\dagger}$. To avoid the catastrophic inconsistencies of quantum mechanical generalized theories, the resulting irreversible formulation has to be reformulated in terms of geno-unitary transformations, Eqs. (11), p. 259 of [3-9].

It should be noted that the iso-Schroedinger representation of monograph [39] is incomplete due to Santilli's lack of knowledge in 1983 of the isotopic completion of the Newton-Leibnitz differential calculus, that was eventually identified in 1996, as reviewed in Chapter 4.

3.6. Lie-admissible generalization of Galileo's relativity

One of the majestic features of quantum mechanics is its prediction of the same numeric values under the same conditions at different times. This feature has justly established the validity of the experimental verifications of quantum mechanics for the condition of its original conception, that for a reversible system of point-like particles in a vacuum. The feature is notoriously due to the invariance of the systems considered under the Galilean and Lorentz symmetries for non-relativistic and relativistic speeds, respectively (Fig. 3.4).

In Santilli's view, no EPR completion of quantum mechanics will resist the test of time unless the new mechanics preserve the indicated feature. Given the importance of the problem, Santilli presented in the literature under the DOE grants the foundations of the following four inequivalent completions of the Galilean symmetry and relativity:

1) The classical Lie-admissible completion for non-conservative NSA systems whose invariance characterizes the time rate of variation of physical quantities rather than the inapplicable conservation law [3-4] [3-6] [3-7].

2) The operator Lie-admissible completions for non-conservative NSA systems [3-5] [3-6] [3-7].

3) The classical Lie-isotopic completions for closed-isolated reversible systems of extended particles with Hamiltonian and non-Hamiltonian internal forces, in which case the invariance characterizes the conventional Galilean total conservation laws [3-9].

4) The operator Lie-isotopic completions for closed-isolated systems of extended particles with Hamiltonian and non-Hamiltonian internal forces [3-9].

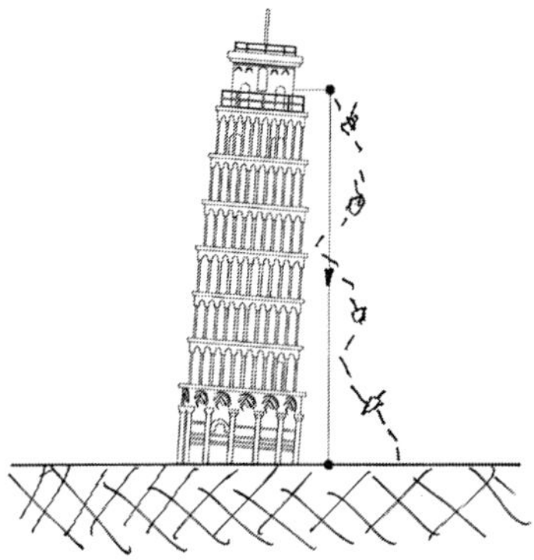

Figure 3.4. By recalling that numerical predictions must be invariant over time to have valid experimental verifications, Santilli initiated in Refs. [3-1]- [3-9] systematic studies for the Lie-admissible generalization of Galileo's relativity for contact non-potential interactions at the classical, here illustrated with a leaf falling from the Pisa tower, and operator levels, illustrated by a proton moving within a star.

During the same period, 1978-1983, Santilli published papers [3-24] [3-25], subsequently studied extensively in various papers and monographs to be reviewed in the subsequent chapters, on the Lie-isotopic completions of the Lorentz symmetry and special relativity for stable, thus reversible systems of extended particles with Hamiltonian and non-Hamiltonian internal forces represented by the Santillian $S = \text{Diag.}(a, b, c, d)\text{Exp}\{-f(t, \psi^*/\psi, \dots)\} \ll 1$, $f>0$, such as stable nuclei and molecules, with local speed of light within physical media $C = c/d$ whose density d is bigger than one, $C < c$, within classical media, e.g., $d = 3/2$ for water, and smaller than one, $C > c$ for hadronic media [3-24]. As we shall see in subsequent chapters, the Lie-isotopic and Lie-admissible completion of the Lorentz symmetry and relativity appear to be fundamental for the possible solution of our environmental problems.

3.7. Lie-admissible connection between mechanics and thermodynamics

As part of his Ph. D. research in the mid-1960's (Sect. 3.1), Santilli studied the connection between mechanics and thermodynamics through the intermediate step of generalized statistical mechanics. By recalling that the characteristics Q of quantum mechanical particles obey Heisenberg's time evolution $idQ/dt = [Q, H] = Q H - H Q$ with conserved energy H, $idH/dt = 0$, against the general non-conservation of thermodynamics, Santilli proved the total disconnection between quantum mechanics and the reality of our macroscopic world with the proof of the following theorem also intended in support of the EPR argument on the lack of completion of quantum mechanics [1-1]:

THEOREM 3.6.1: A macroscopic irreversible thermodynamical system cannot be consistently reduced to a finite number of conservative quantum mechanical particles, and vice versa. A finite number of conservative quantum mechanical particles cannot yield a macroscopic, irreversible, thermodynamical system under the correspondence or other principles.

When studying in the late 1970's irreversible processes at Harvard University under DOE support, Santilli re-examined with his associates [3-20] the earlier studies on the connection between mechanics and thermodynamics by pointing out that the Lie-admissible branch of hadronic mechanics would resolve the problematic aspects identified by Theorem 3.6.1 because macroscopic irreversible thermodynamical systems can indeed be consistently decomposed into a finite number of irreversible particles verifying Lie-admissible dynamical laws and symmetries. The topic was extensively studied at the various meetings under DOE support [3-10] – [3-16].

In essence, with the sole exception of the electron and the proton all particles are naturally unstable, thus irreversible. Consequently, irreversibility appears to originate at the most elementary level of nature (see the titles of monographs [3-6] [3-7]). For this reason, Santilli printed on the front page of the first issue of the *Hadronic Journal* of April 1978 a picture depicting the clear irreversibility of the spontaneous decay of the K-meson, and said picture has been kept as visual evidence of the origin of irreversibility on the front page of the *Hadronic Journal* in the subsequent forty-six years of subsequent regular publication.

3.8. The weakening of Heisenberg's uncertainty principle

The 1981 DOE recorder paper [3-18] indicates that in the transition from point particles in a vacuum under long-range interactions to extended particles in deep entanglement under long-range and contact interactions, a weakening of Heisenberg's uncertainty principle along with Einstein's view [1-1] is inevitable because (Fig. 3.5).

1) On conceptual grounds, the standard deviations of coordinates Dr* and of momenta Dp* for protons and neutrons constrained in the interior of nuclei by strong interactions have to be smaller than the corresponding standard deviations Dr and Dp of electrons orbiting in vacuum around their nuclei.

2) On technical grounds, the representation of the extended character of protons and neutron and their non-Hamiltonian interactions in a nuclear structure via non-unitary transformations characterized by the Santillian S = Diag. (a, b, c, d)Exp{-f(t, ψ^*/ψ, ...)} << 1, f>0, implies a corresponding non-unitary completion of Heisenberg's uncertainty principle with the inevitable weakening of the standard deviations Dr and Dp due to the value S << 1.

The proof, that along Einstein's view, the new standard deviations Dr* and Dp* are smaller than their quantum mechanical counterparts Dr and Dp required a decade of studies on the applicable mathematics and an additional decade of applications and experimental verifications of hadronic mechanics to see that the gene unit generated by the non-unitary transformation XY^+ = I> has a progressively small value with the increase of the hadronic density d, the full recovering of Einstein's determinism being achieved at the value I> = 0 representing gravitational collapse (see Chapter 6 for an outline).

3.9. Verification of hadronic mechanics with the structure of mesons and leptons

3.9.1. Problematic aspects of quark theories

During the period 1978 to 1983, against the teaching by the history of science, the mainstream scientific community was developing one single model for the classification and structure of the so-called 'elementary' particles, that of the SU(3) Lie algebra, whose multiplets allowed a meaningful classification of particles into families, while its fundamental 3-dimensional representation was assumed to characterize the actual, physical, point-like constituents of particles, called quarks, under the exact validity for both, the exterior and interior problems, of relativistic quantum mechanics and special relativity.

At the indicated time, Santilli was under contractual obligation with the U. S. Department of Energy to represent the synthesis of the neutron from the proton and the electron in the core of stars and use the results to search for possible new clean nuclear energies. The conflict between the development of the quark model and Santilli's contractual obligation with the DOE was evident on several grounds, such as [3-22]:

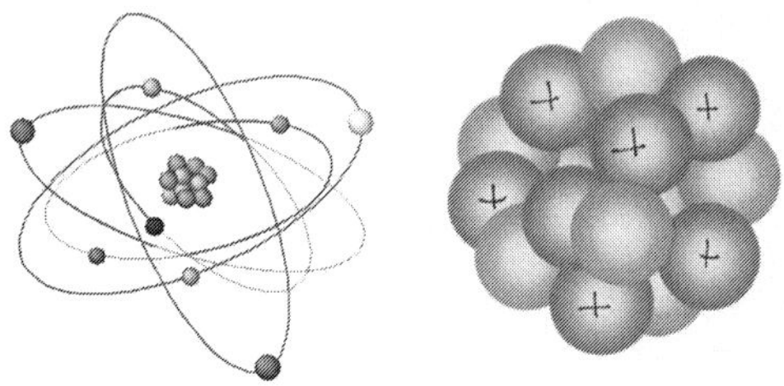

Figure 3.5. In early August 1980, Santilli requested the DOE Officer David Peaslee the transmission to the White House of the nuclear physicist President Jimmy Carter (Section 3-1) paper [3-27] and the preprint at that time of paper [3-22] with the indication that the expected weakening of Heisenberg's uncertainty principle under strong nuclear interactions may trigger a scientific renaissance due to underlining advances in all scientific fields.

1) The assumption that a mathematical representation of the SU(3) characterizes the actual physical constituents of the neutron and other particles was not suitable, as it is the case to this day for a quantitative representation of the neutron synthesis from the proton and the electron and for recycling radioactive nuclear waste and to seek much needed new clean nuclear energies.

2) The assumption that quarks are the actual constituents of the neutron implies that, at the time of the neutron synthesis, the permanently stable proton and electron must disappear and be replaced by the hypothetical quarks. Additionally, at the subsequent time of the neutron spontaneous decay, quarks have to disappear and be somehow replaced by the proton and the electron.

3) Since quarks do not exist in physical reality, they must be permanently and rigorously confined in the interior of particles. However, one of Santilli's graduate students at Harvard University proved a theorem according to which, under the assumption by the quark theory that quantum mechanics and related Hilbert spaces are exactly valid inside and outside

particles, quarks have a finite probability of tunneling free in the outside, against clear experimental evidence that no free quark has ever been detected to the extreme energy collisions of current particle accelerators.

4) Santilli could find no scientific means of representing the gravity of particles whose constituents are quarks because quarks have no physical masses. The sole scientific possibility for masses to be the origin of gravity is that they are defined by the Casimir invariant of the Poincare symmetry, while quarks violate such a basic symmetry due to their peculiar features, such as fractional charges.

5) While accepting the SU(3) classification of particles, Santilli's most dominant reason for rejecting quarks as physical particles was their need for the full interior validity of relativistic quantum mechanics and special relativity. The structural problem of particles whose radius is 1 fm = 10^{-13} cm is characterized by extended wave packets in the condition of total mutual penetration/entanglement, with inevitable, ensuing, non-local, non-linear, and non-potential interactions under which 20th centuries theories cannot be properly formulated.

To illustrate the validity and capability of hadronic mechanics, Santilli decided to complement his research at Harvard University under DOE support with a new particle model resolving the above insufficiencies 1 to 5 of quark models. The new hadronic model is presented in detail in memoir [3-5], particularly in Tables 5.1 to 5.5, pages 827-882, via the implementation of sequential rules that can be summarized as follows for all unstable mesons, leptons, and baryons, here referred to as 'particles':

RULE 3.9.1: All naturally unstable particles are composite, i.e., bound state of smaller constituents, as a necessary condition to admit an internal mechanism responsible for their decay.

RULE 3.9.2: The most plausible, actual, physical, detectable constituents or particles are those produced free in the spontaneous decays, which decays are generally those with the lowest mode.

RULE 3.9.3: Said constituents should identify the spontaneous decay mechanism and allow a numeric representation of the particle mean lives. The most plausible consequence of this rule is that more elementary particles and their antiparticles give the constituents. Spontaneous decays are explained by their destruction, while mean lives are reduced to the value of their kinetic energy or tangential speeds.

RULE 3.9.4: Due to the constituent's kinetic energy, the conservation of the energy implies that the sum of the masses of the constituents is smaller than that of the resulting particle. Consequently, bound states characterizing all unstable particles have a *mass excess* for which quantum mechanics is inapplicable.

RULE 3.9.5: The geno-Schroedinger equation H(r, p)>|ψ(t, r)> = H(r, p) S(t, ψ*/ψ, ...) |ψ(t, r)> = E> |ψ(t, r)> (Sect. 3.3) and its geno-Galilean symmetry (Sect. 3. 6) with the Santillian having simple realizations of the ty8pe $S = d \exp(- kV_h/V_c) \approx 1 - kV_h/V_c$ where V_h is the Hulten potential in representation of the strong interactions and V_c is the Coulomb potential, should provide a numerically exact and time invariant representation of all characteristics of the particle considered while smoothly recovering quantum mechanics for all mutual distances bigger than the particle size (note that the Hulten potential appears in the exponent of the Santillian, and therefore, it does not represent potential interactions).

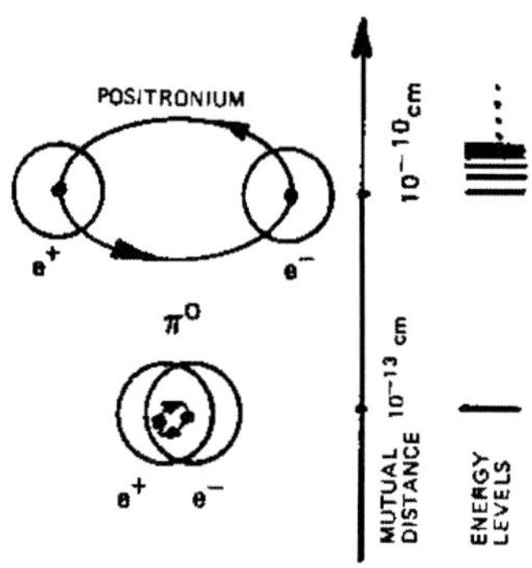

Figure 3.6 Santilli showed in memoir [3-5] that, unlike the case for quark models Section 3.9.1), the hadronic structure model of the π0 meson as a compressed positronium represents all characteristics of the particle, including a mechanism explaining its spontaneous decay.

RULE 3.9.6: The contact, zero-range interactions among the constituents should yield a null contribution to the total energy of the hadronic bound state, as expected by their main definition of not admitting a potential energy.

RULE 3.9.7: The resulting hadronic structure model of particles should be compatible with quantum mechanical SU(3) classifications of particles into families.

3.9.2. Santilli's structure model of mesons with physical constituents

The main characteristics of the π^0 meson are the following:

1) Mass m =134.967 M eV.

2) Mean-life $\tau = 0.828 \times 10^{-16}$ s.

3) Charge radius $R = 10^{-13}$ cm.

4) Null charge and spin.

5) Null electric and magnetic moments.

6) Negative parity.

7) Primary decay $\pi^0 \rightarrow \gamma + \gamma$,

By using Rules 3.9.1- 3.9.3, Santilli assumed that the physical constituents of the π^0 are given by an electron and a positron, or more precisely, that the π^0 is compressed positronium down to the charge radius of 1 fm of hadronic mechanics (hm), $\pi^0 = (e^-, e^+)_{hm}$. This assumption was made because the electron and the positron are actual physical particles; their annihilation into photons explains the mechanism of the spontaneous decay as well as their primary decay mode; their kinetic energy allows the representation of the π^0 mean life (Fig, 3.6).

All characteristics of the π^0 were represented exactly and invariantly via the geno-equation of Rule 3.9.5, Eqs. (5.1.14), p. 836 of [3-5], under the iso-Galilean symmetry [3-9] at the non-relativistic level and the iso-Lorentz symmetry [3-25] at the relativistic level. The study included the detailed analytic solution of the underlying differential equations, including their boundary condition (pages 836 to 841 of [3-5]), which continues to be used in various papers today.

The lack of characterization of any energy by the Santillian (Rule 3.9.6) is presented in Eq. (5.1.36), p. 841. It is technically expressed by the fact that zero is the sole permissible value of the finite energy spectrum of the Hulten potential following the numerical solution (5.1.42), p. 840 of [3-5].

The hadronic π^0 model here considered also indicated that, as a necessary condition to achieve a solution of the background differential equations, the transition from the positronium at large mutual distances to the π^0 at 1fm mutual distance, the intrinsic characteristics of the electron e*- and the positron er*+ are mutated into a form technically identified by the transition from the Galilean to the iso-Galilean or from the Lorentz to the iso-Lorentz symmetry. The mutated electrons were called *eletons* in memoir [3-5] (a name recently replaced by *iso-electron*), resulting in the structure $\pi^0 = (e*^-, e*^+)hm$.

The mutation $(e^-, e^+)qm \rightarrow = (e*^-, e*^+)hm$ allowed the following *geometric* representation of the mass excess in the p π^0 synthesis from the positronium. Since the mass of the π^0 is 134.976 MeV and the mass of the electron/positron is 0.511 MeV, the π^0 mass excess Δm is given by Δm = 134.976 - 2 x 0.511 = 133.954 MeV, corresponding to the iso-renormalized mass of the electron and of the positron m* = 66.977 MeV. From the energy iso-equivalence [3-25] mC^2 = mc^2/d^2, we then obtain the density of the π^0 d = (0.511/66.977)^1/2 = 0.007, which indicates a very low energy density of the π^0as expected from the π^0constituents being the lightest particles existing in nature. By Rule 3.9.4, a mass excess occurs to synthesize *all* unstable particles from lighter constituents, including the neutron from a proton and an electron in the core of stars.

Note that the use in the geno-Schrödinger equation of the iso-renormalized mass of the electron m = 0.511 MeV → m* = 66.977 MeV reconstructs the conventional *mass defect* caused by Coulomb interactions, by therefore allowing a solution of the differential equations that would be otherwise impossible (technically due to the inconsistency of the indicial equation of the Schrödinger equation under the condition of a mass excess). On *physical* grounds, the excess energy the electron and the positron need to create the π^0 is carried by their kinetic energy E = m* - m = 66.977 - 0.511 MeV = 66.766 MeV, which is quite big, as expected.

The hadronic structural model of heavier neutral mesons was achieved in Table 5.1 of [3-5] via a bootstrapping of the π0model, such as K0 = (π*˜0, π*˜0)hm, KS = (π*˜−, π*˜+)hm, KL = (K*˜−, π*˜+)hm and they all exhibited a mass excess according to Rule 3.94. The hadronic structural model of the remaining four charged mesons, which were needed to recover the meson octet of the SU(3) classification per Rule 3.9.7, $\pi^\pm = (\pi^{*0}, e^{*\pm})_{hm}$, $K^\pm = (\pi^{*0}, \pi^{*\pm})_{hm}$, could not be studied in memoir [3-5] due to the prior need for in-depth studies on the isotopies of the SU(2)-spin.

3.9.3. Santilli's structure model of leptons with physical constituents

The main characteristics of the mu lepton $\mu^\pm$ are the following:

1) Rest energy 105, 658 M eV;

2) Charge radius $R = 10^{-13}$ cm;

3) Mean-life $\tau = 2.19703 \times 10^{-6}$ s;

4) Spin 1/2 and elementary charge;

5) Spontaneous decay $\mu^\pm \rightarrow e^- + e^\pm + e^+$.

In Table 5.2, pages 849-868 of memoir [3-5], Santilli proved that the geno-Schrödinger equation (5.2.4) for the model $\mu^\pm = (e^{*-}, e^{*\pm}, e^{*+})_{hm}$ represented *all* characteristics of the particle considered. Repeating the analytic solution of the new differential equations for the new boundary conditions confirmed this result.

Note that the non-potential interactions of hadronic mechanics are applicable for representing all cases in which we have a deep mutual overlapping/entanglement of the wave packets of particles beginning with strong nuclear interactions but also including the deep wave overlapping of the electrons in the muon structure.

The hadronic structure model of the tau lepton was not considered in detail in Table 5.2 due to uncertainties in its spontaneous decay and neutrino emissions, whose treatment would require a hadronic generalization of quantum electrodynamics.

Note that the non-potential interactions of hadronic mechanics are applicable for representing all cases in which we have a deep mutual overlapping/entanglement of the wave packets of particles beginning with strong nuclear interactions but also including the deep wave overlapping of the electrons in the muon structure.

Note that none of the hadronic structure models of mesons and leptons presented in the memoir [3-5] is possible under the validity in their interior of Heisenberg's uncertainty principle, while the same models are possible under its generalization for strong interactions [3-21] (Fig. 3.7).

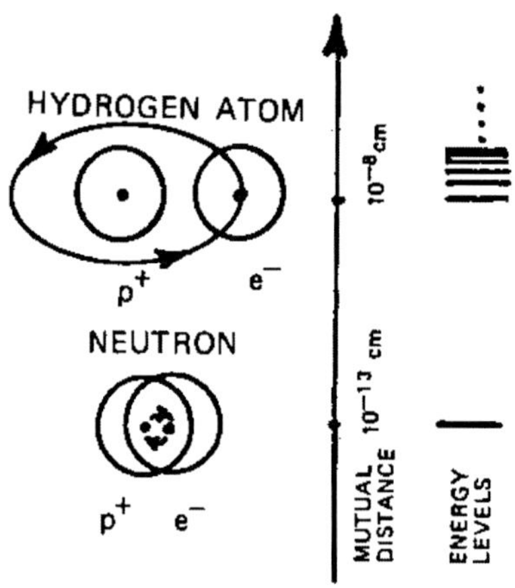

Figure 3.7. Following the exact and invariant representation of all meson and lepton data, Santilli concluded memoir [3-5] with the indication that the achievement of corresponding results for baryons, beginning with the neutron, required the full development of the Lie-Santilli isotopic theory [3-9], with particular reference to the hadronic generalization of the atomic notion of spin reviewed in the following chapters.

In closing, let us quote the view expressed by Sir Karl Popper [1-2], according to his memoir [3.5]

"... seem to augur a return to sanity: to that realism and objectivism for which Einstein stood, and which had been abandoned by great physicists."

Epilogue

In this Chapter 3 and quoted references we outlined the research performed in 1978-1981 at Harvard University under 1978-1981 DOE grants ER-78-S-02- 47420.A000, AS02-78ER04742, DE-ACO2-80ER10651 and the research peformed in 1981-1983 at the Institute for Basic Research (IBR), R. M. Santilli president, under DOE grants DE-ACO2- 80ER-10651.A001, DE-ACO2-90ER10651.A002.At the initiation of the IBR grants, Ms.

Higashi fom the Chicago DOE office called Santilli under eyewitnesses to state that, contrary to the approved budget of the grants, no administrative/accounting funds were allowed. Santilli then had no other choice than doing all administrative works while maintaining the primary records at Bay Bank Harvard Trust and personal records at his accountant. On July 1983 the DOE Officer Gillespie visited the IBR, then located in Cambridge, MA, to audit the grants. Santilli provided evidence that his personal copies of all issued checks had been lost in a two alarm fire reported by local news at the office of his accountant, but suggested to Mr.Gillespie to base his audit of IBR grants on the original copies of all checks available at Bay bank Harvard Trust. Following an inspection of the volume of research done by scientists the world over under said grants, Gillespie stated under eyewitnesses that "There is no need to use the records of Bay Bank Harvard trust" and left the IBR office. Hence, to our knowledge, the indicated IBR grants from the DOE remained unaudited.

References of Chapter 3

[1] Santilli, R. M.: Embedding of Lie-algebras into Lie-admissible algebras, *Nuovo* Cimento **1**, 570-585 (1967), http://www.santilli-foundation.org/docs/Santilli-54.pdf

[2] Santilli, R. M.: Dissipativity and Lie-admissible algebras, *Meccanica* **4**, 3-11 (1969), http://www.santilli-foundation.org/docs/meccanica-1969.pdf

[3] Santilli, R. M.: An introduction to Lie-admissible algebras, *Suppl. Nuovo Cimento* **6**, 1225 (1968), www.santilli-foundation.org/docs/Santilli-501.pdf

[4] Santilli, R. M.: On a possible Lie-admissible covering of Galilei's relativity in Newtonian mechanics for nonconservative and Galilei form-non-invariant systems, *Hadronic Journal* **1**, 223-423 (1978), http://www.santilli-foundation.org/docs/Santilli-58.pdf

[5] Santilli, R. M.: Need of subjecting to an experimental verification the validity within a hadron of Einstein special relativity and Pauli exclusion principle, *Hadronic Journal* 1, 574-901 (1978), http://www.santilli-foundation.org/docs/santilli-73.pdf

[6] Santilli, R. M.: *Lie-Admissible Approach to the Hadronic Structure*, Vol. I International Academic Press, (1978), http://www.santilli-foundation.org/docs/santilli-71.pdf

[7] Santilli, R. M.: *Lie-Admissible Approach to the Hadronic Structure*, Vol. II International Academic Press, (1982), http://www.santilli-foundation.org/docs/santilli-72.pdf

[8] Santilli, R. M.: *Foundation of Theoretical Mechanics,* Springer-Verlag, Heidelberg, Germany, Vol. I (1978) *The Inverse Problem in Newtonian Mechanics,* http://www.santilli-foundation.org/docs/Santilli-209.pdf
[9] Santilli, R. M.: *Foundation of Theoretical Mechanics*, Springer-Verlag, Heidelberg, Germany, Vol. II (1983) *Birkhoffian Generalization of Hamiltonian Mechanics,* http://www.santilli-foundation.org/docs/santilli-69.pdf
[10] Myung, H. *C.* and Santilli, R. M., Editors, Proceedings of the Second Workshop on Lie-Admissible formulations, Part A: Review Papers, *Hadronic Journal* 2, No. 6 (1979), http://www.santilli-foundation.org/docs/hj-2-6-1979.pdf. Part B: Research Papers, *Hadronic Journal* 3, No. 1 (1979), http://www.santilli-foundation.org/docs/hj-3-1-979.pdf
[11] Myung, H. C. and Santilli, R. M., Editors, Proceedings of the Third Workshop on Lie-Admissible Formulations, Part A: Mathematics, *Hadronic Journal* 4, 183-607 (1981), http://www.santilli-foundation.org/docs/hj-4-2-1981.pdf, Part B: Theoretical Physics, *Hadronic Journal* 4, 608-1165 (1981), http://www.santilli-foundation.org/docs/hj-4-3-1981.pdf, Part C: Experimental Physics, and Bibliography, *Hadronic Journal* **4**, 608-1165 (1981),
http://www.santilli-foundation.org/docs/hj-4-4-1981.pdf
[12] Arenas, T., Fronteau, J. and Santilli, R. M., Editors, *Proceedings of the First International Conference on Nonpotential Interactions and their Lie-Admissible Treatment,* Part A: Invited Papers, *Hadronic Journal* **5**, No. 2 (1982), http://www.santilli-foundation.org/docs/hj-5-2-1982.pdf, Part B: Invited Papers, *Hadronic Journal* **5**, No. 3 (1982),
http://www.santilli-foundation.org/docs/hj-5-3-1982.pdf,
Part C: Contributed Papers, *Hadronic Journal* **5**, No. 4 (1982),
http://www.santilli-foundation.org/docs/hj-8-4-1982.pdf,
Part D: Contributed Papers, *Hadronic Journal* **5**, No. 5 (1982),
http://www.santilli-foundation.org/docs/hj-5-5-1982.pdf
[13] Myung, H. C. and Santilli, R. M., Editors, Proceedings of the First Workshop on Hadronic Mechanics, *Hadronic Journal* **6**, No. 6 (1983),
http://www.santilli-foundation.org/docs/hj-6-6-1983.pdf
[14] Myung, H. C. and Santilli, R. M., Editors, Proceedings of the Second Workshop on Hadronic Mechanics, Vol. 1: *Hadronic Journal* **7,** No. 5 (1984) http://www.santilli-foundation.org/docs/2nd-work-HM-I.pdf, Vol. II: *Hadronic Journal* **6**, No. 6 (1984), http://www.santilli-foundation.org/docs/2nd-work-HM-II.pdf

[15] Myung, H. C., Okubo, S. and Santilli, R. M., Editors, Applications of Lie-admissible Algebras in Physics, Reprint series, Hadronic Press (1984), Vol. I: Originating papers, http://www.santilli-foundation.org/docs/hj-7-5-1984.pdf, Vol. II: Initial developments, Hyperlink to be provided, Vol. III: Direct universality of Lie-admissible algebras, http://www.santilli-foundation.org/docs/hj-7-6-1984.pdf

[16] Schoeber, A., Editor, Irreversibility and Non-potentiality in Statistical Mechanics, Hadronic Press (1984), http://www.santilli-foundation.org/docs/Santilli-110.pdf

[17] Tomber, M. L. et al., Bibliography and indes in nonassociative algebras, Combined Volumes I, II, III, Hadronic press (1984), http://www.santilli-foundation.org/docs/Tomber-bibliography.pdf

[18] Santilli, R. M.: Addendum to: On a possible Lie-admissible covering of Galilei relativity in Newtonian mechanics for nonconservative and Galilei form noninvariant systems, *Hadronic Journal* **4**, 1279-1342 (1978), Hyperlink ti be provided

[19] Santilli, R. M.: Initiation of the representation theory of Lie-admissible algebras of operators on bimodular Hilbert spaces, *Hadronic Journal*, 440-467 (1979), http://www.santilli-foundation.org/docs/santilli-1978-paper.pdf

[20] Fronteau, J., Tellez-Arenas, A. and Santilli, R. M.: Lie-admissible structure of statistical mechanics, *Hadronic Journal* **3**, 130-176 (1979), http://www.santilli-foundation.org/docs/arenas-fronteau-santilli-1981.pdf

[21] Santilli, R. M.: Generalization of Heisenberg's uncertainty principle for strong interactions, *Hadronic Journal* **4**, 642 (1981), https://www.osti.gov/biblio/6648552

[22] Santilli, R. M.: An intriguing legacy of Einstein, Fermi, Jordan, and others: The possible invalidation of quark conjectures, *Found. of Phys.* **11**, 384-396 (1981), www.santilli-foundation.org/docs/Santilli-36.pdf

[23] Myung, H. C. and Santilli, R. M.: Modular-isotopic Hilbert space formulation of the exterior strong problem, Hadronic Journal **5**, 1277-1366 (1982), www.santilli-foundation.org/docs/Santilli-201.pdf

[24] Santilli, R. M.: Can strong interactions accelerate particles faster than the speed of light? *Lettere Nuovo Cimento* **33**, 145-153 (1982), https://www.santilli-foundation.org/docs/Santilli-102.pdf

[25] Santilli, R. M.: Lie-isotopic Lifting of Special Relativity for Extended Deformable Particles, *Lettere Nuovo Cimento* **37**, 545-555 (1983), www.santilli-foundation.org/docs/Santilli-50.pdf

[26] Scientia Editorial Board: Progressive Recovering of Einstein's Determinism under Strong Interactions, Scientia November 2024, https://doi.org/10.33548/SCIENTIA1115
[27] Scientia Editorial Board: Completing Quantum Mechanics: Hadronic Mechanics and its Potential for Clean Energy, Scientia December 2024, https://doi.org/10.33548/SCIENTIA1133
[28] Muktibodh, A: Santilli's recovering of Einstein's determinism, *Progress in Physics* **20,** 26-34 (2024), http://www.eprdebates.org/docs/muktibodh-2024.pdf
[29] F. Inglese, F. et al: Lie-admissible irreversible biological entanglements and their apparent initiation of hadronic medicine *Ratio Mathematica* **54**, 1-32 (2025), hztp://www.santilli-foundation.org/docs/hadronic-medicine.pdf

CHAPTER 4

CLASSIFICATION OF HADRONIC MATHEMATICS AND MECHANICS

4.1. The challenges for building new mathematics

In view of the above foundational aspects, Santilli has expressed various times in his writings that Physics experiments produce numbers, that to have scientific value, must be elements of numeric fields whose axioms have been identified by great mathematicians such as André Weil, Carl Friedrich Gauss, Pierre de Fermat, and others. 20th century applied mathematics is a magnificent body of mutually compatible methods, including functional analysis, Lie's theory, geometry, topology and other methods all defined on a numeric field.

Given the above foundational aspects, Santilli has expressed various times in his writings that a physical theory cannot be new unless it is based on new mathematics, and a mathematics cannot be new unless it is based on new numbers.

Following the inception reviewed in Chapter 3 of the new Lie-isotopic and Lie-admissible mathematics, the priority for their development was identifying the respective numeric fields in which they are defined. The power of the axioms of a numeric field soon came to full light and so did the challenges for building new mathematics.

A numeric field can be essentially reduced to a set $F(n, +, \times, 0, 1)$ of elements n with two binary operations, the addition denoted $nm = n + m$ with additive unit 0 in F for which $n + 0 = 0 + n = n$ for all n in F and the multiplication denoted $n \times m$ with multiplicative unit $1 \in F$ for which $n \times 1 = 1 \times n = n$ for all n in F, both operations satisfying a list of axioms. Even though the sum is not altered in the new Lie-isotopic and Lie-admissible mathematics, it was feared that their generalization of the basic associative product nm with the insertion of the Santillan S, $n \star m = nSm$ (Sect. 3.4) and ensuing necessary generalization of the multiplicative unit $1 \rightarrow I^* = 1/S$, may violate the axioms of a numeric field, with ensuing inability by hadronic mechanics to have valid experimental verifications for Santillians of the type S = Diag. Matrix (a, b, c, d)× exp [-f(t, ψ, ...)], f > 0, for the representation of volume V occupied by extended particles V = Diag. (a, b, c), their density D = d and their new interactions G = exp-f(t, ψ,,...). At the same time, the definition of the new Lie-isotopic and Lie-admissible mathematics on a conventional numeric field F leads to catastrophic inconsistencies in the data elaboration of the experiments, by therefore forcing the new mathematics to be systematically defined in all their branches in terms of the new product n⋆m = nSm with new multiplicative unit I*= 1/S.

Following Santilli's discovery in 1993 [4-1] of the needed new numbers verifying all axioms of a numeric field (reviewed in Section 4-2) and thanks to contributions by numerous mathematicians, virtually all aspects of 20th century applied mathematics were reformulated in such a way to be consistently defined on the new numeric field (see Th. Vougiouklis [4-3] for the extension to hypernumbers).

Despite that, hadronic mechanics did not reach maturity for experimental verifications because of the lack of the crucial invariance of numeric predictions under the same conditions at different times because of the lack of consistent symmetries within the context of the Lie-isotopic theory [3-9]. Following years of relentless searches throughout all branches of the new mathematics, the origin of the lack of invariance was finally identified in the Newton-Leibnitz differential calculus [4-2] because of its locality, that is the sole possibility of formulating differentials and derivatives at isolated points in space. In contrast, the new mathematics had to represent the volume V = Diag. (a, b, c) of extended particles and their new interactions.

Santilli finally achieved in 1996 [4-2] the extension of the 1695 Newton-Leibnitz differential calculus to volumes, which extension is systematically studied in the 1995 volumes [4-4] [4-5] and cannot be reviewed here]. The new differential calculus was proved to allow new, consistent, Lie-isotopic and Lie-admissible space-time symmetries with ensuing invariance over time of the numeric predictions, thus signaling the achievement of maturity of the new mathematics and related new mechanics. Systematic physical applications of hadronic mechanics to experiments were then initiated soon after that and reported in the 2016 volume [4-6] (see Ref. [4-7] for a summary and monographs: [4-8] for a recent overview; [4-9] for applications to chemistry; [4-10] for biology applications; [4-11] for applications to antimatter; [4-12] to [4-17] for independent mathematical studies; [4-18] to [4-21] for independent physical studies); and [2-5] [2-6] for recent general reviews of the new mathematics underlying hadronic mechanics and its applications in physics, chemistry and biology).

4.2 Classification of hadronic mathematics

As indicated in preceding chapters, the primary aim of the EPR completion [1-1] of quantum into hadronic mechanics is to provide a deeper representation of stable, thus time reversal invariant nuclei composed by extended protons and neutrons in condition of mutual overlapping with potential/Hamiltonian and contact/non-Hamiltonian interactions; irreversible nuclear fusions between pairs of said nuclei as well as fuel combustions; complex irreversible multi-valued systems such as DNA; and antimatter according to P. A. M. Dirac's conception of antimatter as having negative energies.

The above systems are so structurally and topologically different that to require specifically constructed mathematics for optimal treatment. In turn, the new mathematics must be formulated on correspondingly different numeric fields to be consistent. In this section, we review the results of a decade-long search of all possible, physically inequivalent new numbers that can be defined via an axiom-preserving completion of the associative product AB of quantum mechanics into the iso-associative product $A \star B = ASB$ of hadronic mechanics and related isounit $I* = 1/S$.

Let F (n, ×, 1) be a numeric field, that is, a conventional set of real, complex, or quaternionic numbers n equipped with the conventional sum n + m and multiplication n m, additive unit 0, and multiplicative unit 1, which set verifies all axioms by Weil, Gauss, Fermat and other founders of number theory.

By ignoring the addition for simplicity, Santilli essentially discovered in Ref. [4-1] that the abstract axioms of a numeric field do not necessarily require that the multiplicative unit be the millenary number 1 with consequential, equally millenary multiplication nm since the multiplicative unit can be any desired quantity I∗ with an arbitrary (non-singular) functional dependence on local variables, $I* = I*(t, r, p, \psi, ...)$, thus not necessarily being $\in F$, provided that:

1) The new multiplicative unit I∗ is invertible. Namely, there exists a quantity S such that $I* = 1/S$;

2) Conventional numbers n are generalized into the new isonumbers n∗ = n I∗;

3) The conventional product nm is generalized into the new form n ⋆ m = nSm.

Ref. [4-1] then proved a theorem according to which the new set F∗(n∗,⋆, I∗) under conditions 1), 2), 3) verifies all axioms of a numeric field, thus being usable for experimental verifications of hadronic mechanics. Additionally, Ref. [4-1] provided the following physically meaningful classification of the ensuing new numbers and related mathematics:

I) **Isonumbers** [4-1] [4-4] [4-14], including isoreal, isocomplex, and isoquaternionic isonumbers n∗ = nI∗ which occur when the multiplicative unit is a positive-definite arbitrary quantity I∗ = 1/S > 0, including ordinary numbers, functions, matrices, integrals, etc. Isonumbers can indeed be used in the experimental verifications of hadronic mechanics because the factor I∗ cancels out the isoeigenvalue equation on a Hilbert space H⋆|ψ⟩ = HS|ψ⟩ = E∗⋆|ψ⟩ = E(1/S)S|ψ⟩ = E|ψ⟩, by therefore resulting in conventional measurable predictions E in F.

II) **Genonumbers** [4-1] [4-4] [4-5], including genoreal, genocomplex, and genoquaternionic genonumbers, which occur when the multiplicative unit Ic is invertible and has complex values with general expression nc = n Ic. Note the compatibility of genonumbers with the axiomatization of the irreversibility of Sect. 3.5.3. Hence, they qualify for the characterization of irreversible processes. This can be seen by recalling that quantum mechanics has attempted to represent irreversible nuclear processes via a complex-valued Hamiltonian Hc with a consequential lack of proper definition of the energy, let alone its loss of Hermiticity/observability, as well as the occurrence of complex-valued eigenvalues, Hc|ψ⟩ = Ec|ψ⟩. By contrast, non-conservations are represented in hadronic mechanics with the forward Santillian S. Consequently, the Hamiltonian H maintains its Hermiticity-observability although with non-conserved real eigenvalues, H ⋆ |ψ⟩ = HS|ψ⟩ = Ec⋆ |ψ⟩ = E|ψ⟩, E ∈ F. Th representation of the time rate of variation of the energy is done va the Heisenberg-Santilli Lie-admissible time evolution (Sect. 3.3) idH/dt = (H, H) = H< H - H > H = HRH - HSH = H(R - S)H which, for the standard realization of the backward and forward Santillians R = 1, S = 1 - F/H reduced to idH/dt = HF, where F is the operator version of the external terms in Lagrange's and Hamilton's equations.

III) **Hypernumbers** [4-3] [4-10] which occur when the hyperunit Ih is characterized by an ordered set of values under the special operations of the hyperstructures, in which case the hyperproduct yields an ordered set of an arbitrary number of values, as desired.

IV) **Isodual isonumbers, isodual genonumber, and isodual hypernumbers** [4-1] [4-11] hereon denoted with the upper index d, which occur when the generalized unit Id is the anti-Hermitean image of the preceding unit, e.g., for isodual isonumbers, we have $Id = -I*\dagger$. As we shall see, this mathematical assumption alone is sufficient to ensure that particles' mass, energy, and other positive characteristics become negative by Dirac's conception of antimatter.

On physical grounds, the primary difficulty in using the new numbers is conceptual rather than technical. A main objective of the new mathematics is the representation of a particle's dimension and shape, which, in a three-dimensional space, is done via the generalized isounit $I* = \text{Diag.}(1/a, 1/b, 1/c)$. Said representation is mathematically consistent if and only if applied to the totality of the formulations, beginning, most importantly, with the redefinition of numbers into isonumbers, which now become volumes, $n* = n\, I* = n\, \text{Diag.}(1/a, 1/b, 1/c)$, thus requiring some time for its familiarity. Similar conceptual difficulties exist for the use of genonumbers and isodual numbers.

In summary, **hadronic mathematics** [4-8] [4-9] [2-5] comprises **isomathematics, genomathematics, hypermathematics, and isodual mathematics**, which can be conceived as step-by-step completions of 20th century applied mathematics into forms admitting the most general possible multiplicative units and related generalized realizations of the associative product at all their possible levels.

never end thanks to the never-ending advances in number theory.

4.3 Rudiments of isomathematics

To render this presentation minimally self-sufficient, let us recall that Lie-isotopic mathematics, also called *Santilli's isomathematics,* is an axiom-preserving completion of 20th-century mathematics that, as such, can be constructed via the following two-step isotopies [4-4]:

STEP 1: Subject every quantity and related operations of 20-th century applied mathematics to the non-unitary transformation $A \to A' = UAU\dagger$,

UU† = 1∗ = 1/S > 0 in such a way that the Santillian S = S(r, p, ψ, ...) = (UU†)−1 = S† = 1/1∗ is a time independent, positive-definite quantity, resulting in [4-2]: the isoproduct AB→U(AB)U† = U′SU′† = A' ⋆ B'; isonumbers n∗ = UnU† = n1*; isofunctions such as the isoexponent of an isooperator, e∗X = =U[exp(X)U† = [exp(XS)]1*= 1*[exp(SX)] and isofunctions in general f∗(r∗,...)=[f(r1∗,...)]1∗; isocoordinates r* = r1* and related isospaces; isodifferential d∗r∗ =Sd[r1∗(r,p,ψ,...)]=dr+rSd1∗ and isoderivative ∂∗f∗(r∗, ...)/∂∗r∗ ≈ 1∗∂f(r1∗)/∂r; Lie-Santilli isoalgebras; isosymmetries; etc.

Figure 4.1. Santilli's discovery of the new iso-, geno-, hyper- and isodual numbers was particularly inspired by Johann Carl Friedrich Gauss (April 30, 1777 – February 28, 1855), one of the founders of the modern number theory, because of his view that the discovery of new mathematics will

STEP 2: To prevent inconsistencies caused by the mixing of different mathematics, the resulting formalism has to be formulated on isospaces with isoproducts A ⋆ B over isofields of isonumbers with isounit 1 in such a way that isomathematics formulated on isospaces over isofields must coincide at the abstract realization free level with 20th century mathematics formulated on conventional spaces over conventional fields. In particular, the non-unitary transformation UU† =1∗≠ 1 has to be rewritten in the isounitary form U = WS1/2, UU† = W ⋆W† = W†⋆W = 1* to coincide with the hypothesis of unitarity at the abstract level. The same should apply to all other operations. The assumption of the specific realization of the isoproduct ASB, S = Diag. (a, b, c, d) > 0 and isounit I* = 1/S implies the projection of isomathematics into a conventional space over a conventional field which, as such, is afflicted by many insufficiencies treated in Chapter 5.

According to rigorously proved mathematical theorems [4-17], the possible existence of axiomatic differences between the so-defined mathematics and isomathematics is evidence of the erroneous construction of the isotopies.

4.4. Rudiments of genomathematics

Let us recall that Lie-admissible mathematics, or *Santilli's genomathematics,* can be constructed via the following three sequential steps:

GENOSTEP 1: Assume all products ordered to the right $A>B$ denote motion forward and all products ordered to the left $A<B$ denote motion backward in time. The above-ordered assumptions decompose the universal isoassociative algebra U^* of Sect. 3.5.3 into the product of two independent isoassociative envelopes $<U \times U>$.

GENOSTEP 2: Subject every quantity of 20th century applied mathematics and related operations to the non-unitary transformation $A' = PAQ^\dagger$ to characterize motion forward in time via two independent non-unitary transformations $PP^\dagger \neq 1$, $QQ^\dagger \neq 1$ selected in such a way that the forward genounit $1> = P1Q> = 1/S>$ characterizes an invertible, time-dependent, generally complex Santillian $S>(t, r, p, \psi, ...) = (P\ Q^\dagger)-1$ with genoproducts $P(AB)Q^\dagger = A'S>B'$. For the motion backward in time, assume the non-unitary transformation $A' = QAP^\dagger$ with backward genounit $<1 = Q1P^\dagger = 1/<S$ and backward genoproduct $Q(AB)P^\dagger = A'<SB' = A' < B'$. Genomathematics occur for the independent repetition of Isostep 1 for the forward and backward motions.

GENOSTEP 3: Reformulate the forward mathematics on forward genospaces over forward genofields beginning with the reformulation of the non-unitary transformation into the forward genounitary form $P = MS>1/2$, $Q = NS>1/2$, $PQ^\dagger = M^\dagger > N = N > M^\dagger = 1>$. Repeat the reformulation for backward motion. The representation of irreversibility is assured by the property $S> \neq < S$.

For brevity, we shall outline hypermathematics and isodual mathematics later on when considering specific applications.

4.5. Classification of hadronic mechanics

An important feature of the studies herein considered is that the simplest possible realization of quantum axioms, such as associativity, distributivity, linearity, locality, potentiality, etc., characterizes the Copenhagen interpretation of quantum mechanics. In contrast, hadronic mechanics is characterized by the most general possible realizations of said axioms depending on the complexity of the considered "elements of reality" [1-1], according to the following classification:

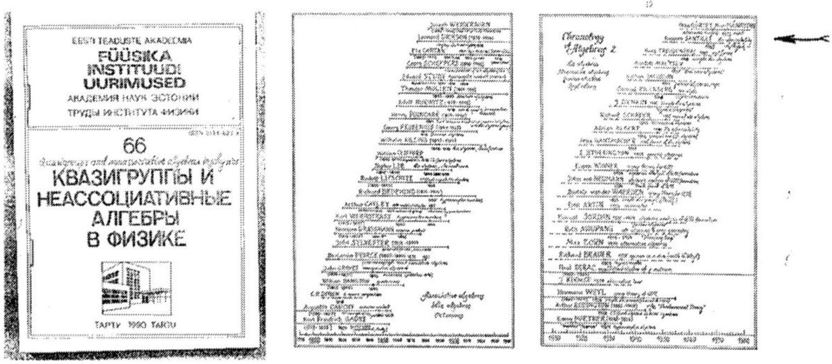

In recognition of the discovery of the new Lie-admissible mathematics, including new numbers, functional analysis, differential calculus, symmetries and relativities, in 1990 the Estonia Academy of Sciences, then under the USSR, nominated Ruggero Santilli among the most illustrious applied mathematicians of all times, jointly with K. F. Gauss, A. Cauchy, W. Hamilton, H. Grassmann, K. Wejerstrass, A. Cayley, R. Dedekind, R. Lipscjiftz, S. Lie, W. Clifford, W. Killing, G. Frobenius, H. Poincare', A. Hurwitz, E. Cartan, L. Dickson, E. Noether, A. Eddington, H. Weyl, P. Dirac, R. Moufang, P. Jordan, E. Artin, J. von Neumann, E. Wigner, F. Gantmaker, A . Albert, R. Schafer, J. Dynkin, S. Eilenberg, N. Jacobsen, A. Maltsev, H. Freudenthal and others.

4.5-I) **Quantum mechanics** for the representation of point-like particles in vacuum under sole potential/Hamiltonian interactions, which occurs as a particular case of branches of all hadronic mechanics when their generalized units acquire the trivial value 1, which generally occurs for particle mutual distances bigger than 1 f m = 10−13 cm. The well-known dynamical equations of quantum mechanics on a Hilbert space with states $|\psi\rangle$ and normalization $\langle\psi|\psi\rangle = 1$ over the field of complex numbers are given by:

1) The Schroedinger representation, which includes the linear momentum $p|\psi\rangle = -i\partial_r |\psi\rangle$, the canonical commutation rules $[r, p]|\psi\rangle = (rp - pr)|\psi\rangle = i1$ and the eigenvalue equations $H(r, p)|\psi\rangle = E|\psi\rangle$.

2) The Heisenberg's representation, which includes the time evolution of an observable A(t) in terms of the Hamiltonian H in its finite unitary form $A(t) = \exp(Hti)A(0)\exp(-itH) = U(t)A(0)U^\dagger$, $UU^\dagger = 1$, with corresponding infinitesimal form for very small periods $idA/dt = [A, H] = AH - HA$.

4.5–II) **Isomechanics** [4-4] [4-5] for the representation of closed conservative systems of mutually overlapping extended nucleons with potential interactions represented with the Hamiltonian H and contact interactions represented with the Santillian S, whose dynamical equations formulated according to isomathematics on an iso-Hilbert space with isostates $|\psi^*\rangle$ and isonormalization $\langle\psi^* | \star |\psi^*\rangle = \langle\psi^* | S |\psi^*\rangle = S > 0$ over the isofield of isonumbers are given by:

1) The Schroedinger-Santilli isorepresentation, [3-9] which includes: the isolinear isomomentum $p^*\star|\psi^*\rangle = -i\partial^*_r|\psi^*\rangle = -i1^*\partial_r|\psi^*\rangle$, $1^* = 1/S$, the isocanonical isocommutaion rules $[r^*, p^*]^*\star|\psi^*\rangle = (r^*\star p^*-p^*\star r^*)\star|\psi^*\rangle = (r^*Sp^*-p^*Sr^*)\, S\, |\psi^*\rangle = i1^*$ and the isoeigenvalue equation $H(r^*, p^*) \star |\psi*\rangle = E^*\, c\, |\psi*\rangle = E\, |\psi^*\rangle$.

2) The Heisenberg-Santilli isorepresentation [3-9] which includes the time evolution of an observable A(t) in terms of the Hamiltonian H and the Santillian S in its finite isounitary form $A(t) = \exp^*(Hti)\star A(t)\star\exp^*(-itH) = \exp(HSti)A(0)\exp(-itSH) = U(t) \star A(0) \star U^\dagger$, $U \star U^\dagger = U^\dagger \star U = 1^*$, with corresponding infinitesimal form for very small periods $idA/dt = [A, H\,]^* = A \star H - H \star A = ASH - HAS$ for the representation of all potential interactions with the Hamiltonian H and all contact interactions with the Santillian S. It should be indicated that the expression i dA.dt = ASH - HSA represents the projection of isomechanics into conventional spaces over conventional fields which is afflicted by insufficiencies treated in Chapter 5.

4.5-III) **Genomechanics** [4-4] [4-5] for the representation of open irreversible processes of deeply entangled/overlapping extended nucleons with potential/Hamiltonian and contact/non-Hamiltonian interactions, whose forward dynamical equations formulated according to genomathematics on a geno-Hilbert space with genostates $|\psi^>\rangle$ and gennonormalization $\langle\psi^>| > |\psi^>\rangle = \langle\psi^>|S^>|\psi^>\rangle = S^>$, where $S^>$ is an invertible, time-dependent, generally complex, forward Santilian over the field of genocomplex genonumbers, are given by:

1) The Schroedinger-Santilli genorepresentation [3-9], which includes the genolinear genomomentum $p^{>} > |\psi^{>}\rangle = -i\partial^{>}_{r}|\psi^{>}\rangle = -i1^{>}\partial_{r}|\psi^{>}\rangle$, $1^{>} = 1/S^{>}$, the genocanonical genocommutaion rules $[r^{>},p^{>}]^{>}>|\psi^{>}\rangle = (r^{>} > p^{>} - p^{>} > r^{>}) > |\psi^{>}\rangle = (r^{>}S^{>}p^{>} - p^{>}S^{>}r^{>})S^{>}|\psi^{>}\rangle = i1^{>}$ and the genoeigenvalue equation $H(r^{>}, p^{>}) > |\psi^{>}\rangle = E^{>} > |\psi^{>}\rangle = E|\psi^{>}\rangle$.

2) The Heisenberg-Santilli genorepresentation [3-9] which includes the time evolution of an observable A(t) in terms of the Hamiltonian H and the Santillian $S^{>}$ in its finite genounittary form $A(t) = \exp^{>}(Hti)>A(t)<\exp^{<}(-itH) = \exp(HS^{>}ti)A(0)\exp(-it^{<}SH) = U>(t) > A(0) > W^{>\dagger}$, $U^{>} > W^{>\dagger} = W^{>\dagger} > U^{>} = 1^{>}$, with corresponding infinitesimal form for very small periods of time $idA/dt = (A,H) = A<H-H > A = A^{<}SH-HS^{>}A = (ATH-HTA)+(AJH+HJA)$, ${}^{<}S=T+J>0$, $S^{>}=T-J>0$, where ${}^{<}S$ is the backward Santillian, with particular form $idA/dt = (A, H) = A<H - H>A = (AH - H A) + F A$ which occurs for ${}^{<}S = 1$ and $S^{>} = 1 - F/H$ by indicating that the representation of tine irreversibility with Lagrange's and Hamilton's external terms F occurs via the addition of the iso-Jordan brackets (A J H + H J A) to the brackets of isomechanics.

4.5-IV) **Hypermechanics** [4-3] [4-10], which is based on Vougiouklis Lie-admissible hyperstructures for the representation of complex biological entities.

4.5-V) **Isodual quantum, iso-, geno- and hyper-mechanics** [4-3] [4-5] [4-11], which are based on the isodual mathematics for the causal representation of antiparticles with negative energy according to P. A. M. Dirac.

References of Chapter 4

[1] Santilli, R. M.: Isonumbers and Genonumbers of Dimensions 1, 2, 4, 8, their Isoduals and Pseudoduals, and 'Hidden Numbers,' of Dimension 3, 5, 6, 7, *Algebras, Groups and Geometries* **10**, 273-295 (1993), http://www.santilli-foundation.org/docs/Santilli-34.pdf

[2] Santilli, R. M.: Nonlocal-Integral Isotopies of Differential Calculus, Mechanics and Geometries, *Rendiconti Circolo Matematico Palermo, Suppl.* **42**, 7-82 (1996), http://www.santilli-foundation.org/docs/Santilli-37.pdf

[3] Vougiouklis, T.: The Santilli 'invasion' in hyperstructures, *Algebras, Groups and Geometries* **28**, 83-104 (2011), http://www.santilli-foundation.org/docs/santilli-invasion.pdf

[4] Santilli, R. M.: *Elements of Hadronic Mechanics,* Ukraine Academy of Sciences, Kiev, Vol. I (1995), Mathematical Foundations, http://www.santilli-foundation.org/docs/Santilli-300.pdf

[5] Santilli, R. M.: *Elements of Hadronic Mechanics,* Ukraine Academy of Sciences, Kiev, Vol. II (1994), Theoretical Foundations, http://www.santilli-foundation.org/docs/Santilli-301.pdf

[6] Santilli, R. M.: *Elements of Hadronic Mechanics*, Ukraine Academy of Sciences, Kiev, Vol. III (2016), Experimental verifications, http://www.santilli-foundation.org/docs/elements-hadronic-mechanics-iii.compressed.pdf

[7] Anderson, R.: Outline of Hadronic Mathematics, Mechanics and Chemistry as Conceived by R. M. Santilli, Special Issue, Foundations of Hadronic Mathematics, Mechanics and Chemistry, *American Journal of Modern* Physics, 1-16 (2016), http://www.santilli-foundation.org/docs/HMMC-2017.pdf

[8] Santilli, R. M.: *Overview of historical and recent verifications of the Einstein-Podolsky-Rosen argument and their applications to physics, chemistry, and biology*, APAV - Accademia Piceno Aprutina dei Velati, Pescara, Italy (2021), http://www.santilli-foundation.org/epr-overview-2021.pdf

[9] Santilli, R. M.: *Foundations of Hadronic Chemistry, with Applications to New Clean Energies and Fuels,* Springer Kluwer Academic Publishers (2001),
http://www.santilli-foundation.org/docs/Santilli-113.pdfAK Russian translation by A.K. Aringazin, http://i-b-r.org/docs/Santilli-Hadronic-Chemistry.pdf

[10] Santilli, R. M.: *Isotopic, Genotopic and Hyperstructural Methods in Theoretical Biology,* Ukraine Academy of Sciences, Kiev (1994), http://www.santilli-foundation.org/docs/santilli-67.pdf

[11] Santilli, R. M.: *Isodual Theory of Antimatter with Application to Antigravity, Grand Unification and the Spacetime Machine*, Springer (2001), http://www.santilli-foundation.org/docs/santilli-79.pdf

[12] Sourlas, D. S. and Tsagas, G. T.: *Mathematical Foundation of the Lie-Santilli Theory*, Ukraine Academy of Sciences (1993), http://www.santilli-foundation.org/docs/santilli-70.pdf

[13] Lohmus, J., Paal, E. and Sorgsepp, L.: *Non-associative Algebras in Physics*, Hadronic Press (1994), http://www.santilli-foundation.org/docs/Lohmus.pdf

[14] Jiang, CX: *Foundations of Santilli Isonumber Theory,* International Academic Press (2001), http://www.i-b-r.org/docs/jiang.pdf

[15] Ganfornina, R. M. F. and Valdes, J. N.: *Fundamentos de la Isotopia de Santilli, International* Academic Press (2001), http://www.i-b-r.org/docs/spanish.pdf English translation: Algebras, Groups and Geometries **32**, 135-308 (2015), http://www.i-b-r.org/docs/Aversa-translation.pdf

[16] Davvaz, B. and Vougiouklis, Th.: *A Walk Through Weak Hyperstructures and Hv-Structures*, World Scientific (2018).

[17] Georgiev, S.: Foundations of IsoDifferential Calculus, Vols. I to VI, Nova Publishers, New York (2014 on). *Iso-Mathematics,* Lambert Academic Publishing (2022), *Introduction to Iso- Euclidean Geometry*, Eliva Press SRL, Republic of Moldova (2024).

[18] Anderson, R.: Editor, *Collected OA Papers on Hadronic Mathematics, Mechanics and Chemistry*, http://www.santilli-foundation.org/docs/HMMC.pdf

[19] Aringazin, A. K., Jannussis, A., Lopez, F., Nishioka, M. and Veljanosky, B.: *Santilli's Lie-Isotopic Generalization of Galilei and Einstein Relativities,* Kostakaris Publishers, Athens, Greece (1991), http://www.santilli-foundation.org/docs/Santilli-108.pdf

[20] Kadeisvili, J. V.: *Santilli's Isotopies of Contemporary Algebras, Geometries and Relativities,* Ukraine Academy of Sciences, Second edition (1997), http://www.santilli-foundation.org/docs/Santilli-60.pdf

[21] Gandzha, I. and Kadeisvili, J. V.: *New Sciences for a New Era: Mathematical, Physical and Chemical Discoveries of Ruggero Maria Santilli*, Sankata Printing Press, Nepal (2011), http://www.santilli-foundation.org/docs/RMS.pdf

CHAPTER 5

PROBLEMATIC ASPECTS OF NON-UNITARY QUANTUM THEORIES

5.1. The majestic axiomatic consistency of quantum mechanics

The integrated methodological body, including number theory, functional analysis, metric spaces, geometries, Lie's theory, space-time symmetries, Hilbert spaces, and other methods, collectively known as 'quantum mechanics,' has justly acquired a historical value because of its important axiomatic features for the operator representation of Newtonian massive points in vacuum under action-at-a-distance, potential interactions with total conserved energy H, among which features we recall: the observability of the total energy thanks to the Hermiticity of the Hamiltonian $\langle\Psi|H\Psi\rangle \equiv \langle\psi H^{\dagger}|\Psi\rangle$; the predictions of the same numeric values under the same conditions at different times. The verification of the superposition principle according to which the total state $|\Psi\rangle$ of a bound state can be decomposed into the sum of the wave functions of the individual constituents $|\Psi\rangle = \Sigma_k \psi_k$, thus allowing the study of the individual constituents of said bound state. These important properties are due to many axiomatic features of quantum mechanics, such as:

5-1-1. **Associativity:** the underlying universal enveloping algebra is associative in the sense that all products of numbers, functions, matrices, etc., are given by the millenary product AB verifying the axiom of associativity (AB)C = A(BC).

5-1-2. **Unitarity:** the time evolution of the theory U(t) verifies the axiom of unitarity $U(t)U(t)^{\dagger} = U(t)^{\dagger}U(t) = I$, as it is the case for Heisenberg's time evolution $A(t) = \exp(Hti)A(0)\exp(Hti)^{\dagger} = \exp(Hti)A(0)\exp(-itH) = U(t)A(0)U(t)^{\dagger}$.

5-1-3. **Linearity:** all dynamical equations solely depend on the first power of the wave function, as is the case for Schroedinger's equation $H(r, p)|\psi\rangle = E|\psi\rangle$.

5-1-4. **Locality:** all physical quantities are solely defined at isolated points in space r and time t, $x = (r, t)$.

5-1-5. **Potentiality**: all admitted interactions have to be derivable from a potential $V(r)$, which is additive to the kinetic energy K in the Hamiltonian $H = K + V$.

Unitarity is particularly important for axiomatic consistency of quantum mechanics because it implies the preservation of Planck's constant $h = 1$ of the basic numeric field $1 \rightarrow U(t)1U(t)\dagger \equiv 1$; the ensuing validity of said field for experimental verifications at all times; the characterization of the infinite class of unitarily equivalent theories according to which all infinitely possible models which can be obtained via a unitary transformation of a given quantum mechanical model are equivalent to each other; and additional important implications.

The following theorem, which is important for the studies reported in this chapter, was proved by Santilli in the 1983 Harvard University monograph [3-9]:

THEOREM 5.1.1: All possible generalizations of quantum mechanics, here referred to all possible generalizations of axiomatic properties 5-1-1 to 5-1-5 when formulated on conventional Hilbert spaces over conventional numeric fields, are characterized by non-unitary transformations of quantum mechanical axioms and dynamical laws.

5.2. Inconsistencies of non-unitary quantum theories

As it is well known, the 1935 historical argument by A. Einstein, B. Podolsky, and N. Rosen that "Quantum mechanics is not a complete theory" [1-1] has stimulated a large number of completions/generalizations of quantum mechanics formulated on a conventional Hilbert space over the field of complex numbers, such as:

1. **Nonlinear theories** [5-1] - [5-3]. We are referring to theories depending on powers of the wave function ψ greater than one. Since the sole variable admitted by quantum mechanics is the Hamiltonian, non-linear theories are generally formulated via equations of the type

$H(r,p,\psi)|\psi\rangle = E|\psi\rangle$ according to which the Hamiltonian H may depend on powers of the wave function ψ.

2. **Nonlocal theories** [5-4] [5-5]. We are referring to theories defined on volumes rather than isolated points, particularly the de Broglie-Bohm non-local and deterministic generalization of quantum mechanics formulated on conventional spaces over conventional fields.

3. **Non-potential theories** [5-6] [5-7]. We are referring to the lie-admissible [5-6] and Lie-isotopic [5-7] generalizations of Heisenberg's equations as originally formulated in 1978 on conventional spaces over conventional fields (Chapter 3).

4. **Generalizations of Heisenberg's uncertainty principle** [5-8] - [5-10]. We are referring to theories implying a generalization of canonical commutation rules used in the definition of quantum uncertainties.

5. **Generalizations of Bell's inequalities** [5-11] - [5-13]. We refer to inequalities admitting classical counterparts, permitting a meaningful study of Einstein's determinism in particle physics. [1-1].

6. **Q-deformations** [5-14] - [5-17]. We are referring to theories that deform the brackets of Lie algebras from their historical form [A, B] = AB - BA to their deformation via a parameter q, (A, B) = A B - q B A while maintaining conventional spaces and numbers.

7. **String theories** [5-18]. We are here referring to a theory reducing our physical reality to infinitesimal vibrating strings defined on multi-dimensional spaces whose interactions produce structures interpreted as everything from particles to large-scale phenomena.

8. **Theories with non-associative envelopes.** Since Lie-admissible algebras U with brackets (A, B) = A<B - B>A = ARB - BSA (Section 3.3) admit Lie algebras as the attached anti-symmetric algebras U^-, it was natural for Santilli to assume in initial works that hadronic mechanics is characterized by non-associative envelopes U. This initial assumption had to be abandoned in the early 1980's because of insurmountable difficulties in constructing symmetries and

relativities in favor of the isoassociative envelope with product A⋆ B = ASB (see Ref. [4-24] for details).

In this chapter we note that, notwithstanding their clear historical character, generalized theories 1 to 7 generally depart from the axiomatic structure of quantum mechanics as a necessary condition to have a new theory. Consequently, said theories are generally afflicted by many problematic aspects studied in detail in Refs. [5-19] - [5-27], including:

5-2-1. **Possible loss of the superposition principle**, which is typically the case for non-linear theories [5-1]. During his tenure at Boston University in the early 1970's, Santilli contacted W. Heisenberg to inquire about his studies in non-linear theories. The epistolary exchange concluded that said theories are mathematically significant. However, they violate the superposition principle, by therefore prohibiting the characterization of the individual constituents of a bound state with said non-linear internal interactions.

5-2-2. **Possible loss of a unitary time evolution**. A sufficient condition for a generalization of quantum mechanics to be nontrivial is to exit from the infinite class of unitarily equivalent theories, which is possible if and only if the time evolution of generalized theories is non-unitary irrespectively of whether explicitly computed or not, $A(t) = W(t)A(0)W^{\dagger}(t)$, $W(t)W^{\dagger}(t) \neq 1$ with ensuing:

i) Lack of preservation over time of the basic unit of the theory $\hbar = 1 \rightarrow 1' = W(t)1W^{\dagger}(t) \neq 1$.

ii) Consequential inapplicability over time of the numeric field on which the theory is defined.

iii) Lack of consistent experimental verifications.

5-2-3. **Possible loss over time of Hermiticity/observability**. Even though at the time $t = t_0$ generalized theories verify the condition of Hermiticity of the conserved Hamiltonian, the same property cannot be consistently defined at a later time due to the loss of the base numeric field, with ensuing loss of the rigorous definition of Hermiticity. Also, the time evolution of several generalized theories has the structure $A(t) = W(t)A(0)W(t)^{-1}$, with consequential loss of observability at all times [5-27].

The lack of consistent definition over time of Hermiticity is typically the case for non-local theories at large, and the de Broglie-Bohm non-local theory in particular, because their non-local broadening of quantum mechanics, such as that via Bohm's hidden variables, require non-local transformation of local quantum models, plus non-unitary deviations from Heisenberg's uncertainty principle. Independently from known criticism, Santilli's comment on the de Broglie-Bohm theory is that it does not allow consistent experimental verifications because conventionally treated non-unitary theories do not preserve over time the unit 1 of the base field, by therefore losing over time the numeric field in which to report measurements. Despite that, non-linear, non-local, and non-potential theories [5-1] - [5-7] have been assumed at the foundation of hadronic mechanics, only subjected to their reformulation via the respective Lie-admissible and Lie-isotopic mathematics.

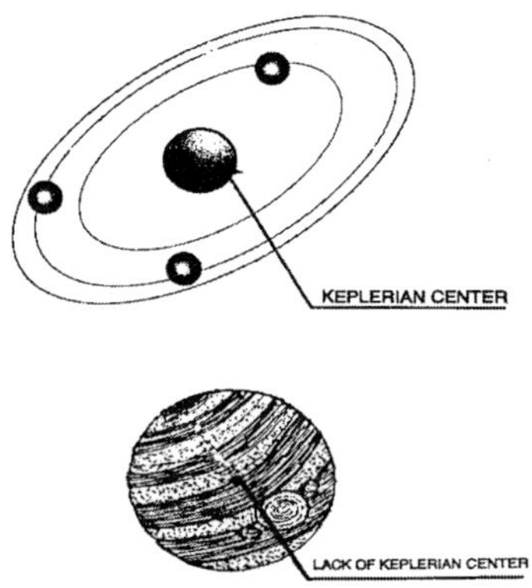

Figure 5.1. A conceptual illustration of the primary difference between exterior and interior dynamical systems: the former (such as planetary or atomic systems) is characterized by a Keplerian center while the latter (such as the structure of a planet or a nucleus) is characterized by its absence, with ensuing validity for the former of the Poincare' symmetry, but need for the latter of its suitable generalization.

The possible loss of Hermiticity/observability over time is also the case for generalizations of Heisenberg's uncertainty principle, string theories, q-deformation, and other theories requiring non-unitary images of quantum mechanical axioms.

Let us review in more detail the general loss of Hermiticity by non-unitary quantum theories and its preservation by hadronic mechanics. Let H be a Hilbert space over a numeric field F with unit 1, states $\psi_1(r)$, $\psi_2(r)$, ... and inner product $\langle\psi_1(r)|\psi_2(r)\rangle \in F$. Then, an operator A is said to be Hermitean (or observable) when it verifies the identity $\langle\psi_1(r)A^\dagger|\psi_2(r)\rangle = \langle\psi_1(r)|A\psi_2(r)\rangle$ $\forall\ \psi_1(r), \psi_2(r) \in H$.

Let $U(r) : H, \rightarrow H^*$, $U(r)U^\dagger(r) \neq 1$ be a non-unitary map of H into H^* with new states $\psi^*(r) = U(r)\psi k(r)U(r)^\dagger$, k = 1, 2, …, and inner product $\langle\psi_1{}^*(r)$ $|\psi_2{}^*(r)\rangle \in F$. Then, the operator A is not necessarily Hermitean/observable on H^* over F because of the possible inequality $\langle\psi_1{}^*(r)A^\dagger \mid \psi_2{}^*(r)\rangle \neq \langle\psi_1{}^*(r)$ $\mid A\psi_2{}^*(r)\rangle\ \forall\ \psi_1{}^*{}_1(r), \psi_2{}^*(r) \in H*$ due to the general non-linear character of the non-unitary nap.

Consider now the isotopic map $U(r) : H \rightarrow H^*$, $U(r)U^\dagger(r\ ) > 0$ of H into H^* with new states $\psi_k{*}(r) = U(r)\psi_k(r)U(r)^\dagger$, k = 1,2,..., and inner product $\langle\psi_1{}^*\ (r) \mid S \mid \psi_2{}^*\ (r)\rangle = \langle\psi_1{}^*\ (r) \mid \star \mid \psi_2{}^*\ (2)\rangle$ on the isofield F^* with isounit $1^* = 1/S = U(r)U(r)^\dagger$, $S = [U(r)U(r)^\dagger]^{-1}$. Then we have the following:

THEOREM 5.1 [4-2]: An operator A which is Hermitean/observable on a Hilbert space H over a numeric field F with multiplicative unit 1 remains Hermitean/observable for all infinitely possible non-unitary isotopic images $U(r)$: $H \rightarrow Hc$ on an isofield F^* with invertible isounit $1^* = UU^\dagger = 1/S$.

PROOF: The new necessary and sufficient condition for Hermiticity/ observability is trivially reducible to the non-unitary map of the original identity $\psi_1{}^*\ (r)\star A^\dagger|\star|\psi_2{}^*\ (r)\rangle = U(r)\langle\psi_1(r)A^\dagger|\psi_2(r)\rangle U(r)^{*\dagger} = \langle\psi_1{}^*\ (r)|\star|A\star\psi_2{}^*$ $(r)\rangle = U(r)\langle\psi_1(r)|A\psi_2(r)\rangle U(r)^\dagger\ \forall\ \psi_1(r), \psi_2(r) \in H$. QED.

In summary, the primary difference between non-unitary quantum theories and hadronic mechanics is that the former theories preserve Planck's quantum $\hbar = 1$ as the unit of the theories at all levels, while hadronic mechanics admits the possible existence of non-local effects under which Planck's quantum of energy is mapped into the generally non-local isounit $1 \rightarrow 1* = U1U\dagger = 1/S$ with consequential isotopic generalization of all aspects of quantum mechanics.

Similarly, Bell's inequalities are crucially dependent on the Lie algebra of Pauli's matrices for the characterization of the spin 1/2 of particles whose generalization is known to be reducible to the non-unitary transformation of quantum models. Therefore, any generalization of Bell's inequalities is afflicted by the same problematic aspects of conventionally formulated non-unitary theories.

5-2-4. **Possible violation of causality**, e.g., the admission of possible solutions in which the effect precedes the cause, which is typically the case for non-unitary time evolution with particular reference to q-deformations [5-14] - [5-17] (see the violation of causality for half-odd-integer eigenvalues of the angular momentum in Ref. [5-28]).

5-2-5. **Possible loss of the associative character of the enveloping algebras**, with con- sequential, inapplicability of Lie's theory, loss of all space-time symmetries, and ensuing lack of preservation over time of numeric predictions [5-21] [5-24].

5.3. Apparent resolution of non-unitary insufficiencies

Recall that quantum mechanics is a formulation based on the simplest possible realization of the associativity of the enveloping algebra and other axioms. A central condition for the correct formulation of hadronic mechanics is that it is an isotopy of quantum mechanics. Namely, hadronic mechanics is characterized by the broadest possible realizations of quantum axioms, therefore implying the identity of quantum and hadronic mechanics at the abstract, realization-free level to such an extent that all quantum and hadronic equations could be expressed via the same symbols, only subjected to different realizations.

This property is illustrated by the abstract identity of the quantum unit 1 and the isounit $1^* = 1/S > 0$, the quantum product AB and the isoproduct $A \star B = ASB$, $S>0$, etc., with the consequential verification by hadronic mechanics of all axiomatic properties of quantum mechanics, such as Hermiticity/observability, predictions of the same numeric values at different times, verification of the superposition principle, etc.

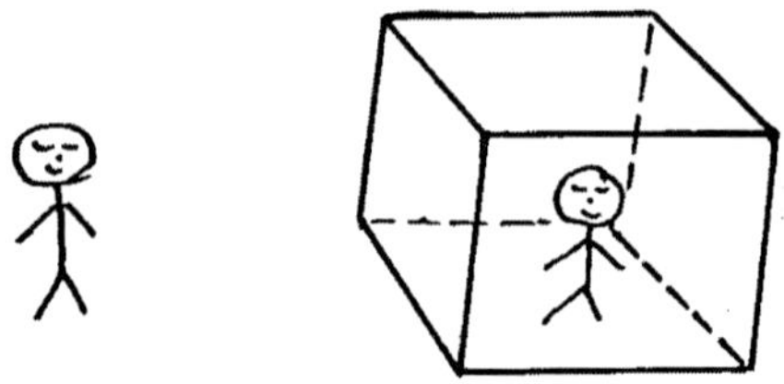

Figure 5.2: Illustration of causal isotopic formulations of non-unitary geometries via the so-called isobox: the interior observer can be in the future or past time with respect to the exterior observer and if the exterior observer sees a cube, the interior observer can see a cathedral.

Consequently, when formulated with the appropriate isomathematics, the notion of isotopy assures the absence of hadronic mechanics of the insufficiencies of non-unitary quantum theories identified in Sect. 5.2. given the reconstruction on isospaces over isofields of all axiomatic properties of quantum mechanics such as (see monograph [5-5] for details):

5-3-1. **Isossociativity,** which refers to the reconstruction of associativity by the isoproduct which $A\star(B\star C) = (A\star B)\star C$.

5-3-2. **Isounitarity, iso-Hermiticity and isoobservability**, which are referred to the reconstruction of unitarity by the time evolution of the theory, $U \star U^{\dagger} = U^{\dagger} \star U = 1^*$ as a result of which operators, such as the Hamiltonian H, that are Hermitean/observable in quantum mechanics xxx = xxx remain Hermitean/observables in hadronic mechanics at all times $\langle\psi^*| \star (U\star H\star U^{\dagger})^{\dagger} \star |\psi^*\rangle = \langle\psi^*| \star |(U\star H\star U^{\dagger}) \star|\psi^*\rangle$.

5-3-3. **Isolinearity, isolocality and isopotentiality,** which are referred to the property that all non-linear, non-local and non-potential terms are embedded in the Santillian S according to the Schroedinger-Santilli isoequation $H(r^*, p^*)\, S(r^*, p^*, \psi^*, ...)\, |\psi*\rangle = H(r^*, p^*) \star |\psi*\rangle = E^* \star|\psi^*\rangle = E\, |\psi^*\rangle$, as a result of which the theory is linear, local and potential on isospaces over isofields, with the ensuing verification of the superposition principle and other quantum properties.

References of Chapter 5

[1] Heisenberg, W.: Nachr. Akad. Wiss. Gottingen IIa, 111 (1953), https://link.springer.com/chapter/10.1007/978-3-642-70079-8_23

[2] Durr, H. P.: On the Nonlinear Spinor Theory of Elementary Particles. In: Urban, P., Editors, Elementary Particle Theories, Acta Physica Austriaca **3**, Springer, Vienna (1966), https://doi.org/10.1007/978-3-7091-5566-0_2

[3] Khalil, H.: Nonlinear systems, Pearson (2001).

[4] Stanford Encyclopedia of Philosophy, 'Bohmian (de Broglie-Bohm) Mechanics' (2021), https://plato.stanford.edu/entries/qm-bohm

[5] Kober, M.: Nonlocal Quantization Principle in Quantum Field Theory and Quantum Gravity, arXiv:1504.03225 (2014), https://arxiv.org/abs/1504.03225

[6] Santilli, R. M.: On a possible Lie-admissible covering of Galilei's relativity in Newtonian mechanics for nonconservative and Galilei form-non-invariant systems, Hadronic Journal **1**, 223-423 (1978), http://www.santilli-foundation.org/docs/Santilli-58.pdf Addendum, Hadronic Journal **1**, 1279-1342 (1978), http://www.santilli-foundation.org/docs/Santilli-58-Addendum.pdf

[7] Santilli, R. M.: Need of subjecting to an experimental verification the validity within a hadron of Einstein special relativity and Pauli exclusion principle, Hadronic Journal **1**, 574-901 (1978), http://www.santilli-foundation.org/docs/santilli-73.pdf

[8] Liang, E. D., Lake, M. J., and Harko, T., Editors: Generalized uncertainty relations: Existing paradigms and new approaches, Special Issue of Frontiers in Astronomy and Space Sciences (2023), www.semanticscholar.org/paper/Editorial%3A-Generalized-uncertainty- relations%3A-and-Lake-Harko/d20ea81adc82ed68453ce5d5f84105e0dc4ba2b4

[9] Li, J-L and Qiao, C-F.: The Generalized Uncertainty Principle, Annalen der Physik **533,** 2000335 (2021), ui.adsabs.harvard.edu/abs/2021AnP...53300335L/abstract

[10] Verdegen, A. and Ziobro, P.: Generalized uncertainty relations, Letters in Mathematical Physics **107**, 659-671 (2017), link.springer.com/article/10.1007/s11005-016-0916-9

[11] Garola, C. and Sozzo, S.: Generalized Observables, Bell's Inequalities and Mixtures in the ESR Model for QM, Found. of Phys. **41**, 424-449, (2011), https://link.springer.com/article/10.1007/s10701-010-9435-1

[12] Karczewski, M.: Avenues to generalizing Bell inequalities, Journal of Physics A **55**, 1-18 (2022), https://iopscience.iop.org/article/10.1088/1751-8121/ac8a28/pdf
[13] Chen, Z-B. et al.: Maximal Violation of Bell's Inequalities for Continuous Variable Systems, Phys. Rev. Lett. **88**, 21-28 (2002), https://doi.org/10.1103/PhysRevLett.88.040406
[14] Biedenharn, L. C.: The quantum group SU(2) and a q-analogue of the boson operators, Journal of Phys. A: **22**, L873–8 (1989).
[15] Macfarlane, A. J.: On q-analogues of the quantum harmonic oscillator and the quantum group SU(2), Journal of Phys. A: **22**, 4581 (1989).
[16] Cardone, F. and Mignani, M.: Energy and geometry, an introduction to deformed special relativity, World Scientific Publishing (2004).
[17] Cardone, F. and Mignani, M.: Deformed spacetime, Springer (2007).
[18] Polchinski, J.: String theory, Cambridge University Press (1998).
[19] Lopez, D. F.: Problematic aspects of q-deformations and their isotopic resolutions, Hadronic Journal **6,** 429-457 (1993), https://www.santilli-foundation.org/docs/Lopez-1993.pdf
[20] Lopez, D. F.: In Symmetry Methods in Physics, **1**, Joint Institute for Nuclear Research, Dubna, Russia (1994), http://www.santilli-foundation.org/docs/Santilli-121.pdf
[21] Jannissis, A., Mignani, R. and Santilli, R. M.: Problematic aspects of Weinberg's nonlinear theory, Annales de la Foundation Louis de Broglie **18**, 371-389 (1993), https://www.santilli-foundation.org/docs/santilli-510.pdf
[22] Santilli, R. M.: Problematic aspects of string theories and their possible resolution, arXiv: physics/9901013 (1999), https://arxiv.org/abs/physics/9901013
[23] Santilli, R. M.: Lie-admissible invariant representation of irreversibility for matter and antimatter at the classical and operator levels, Nuovo Cimento B **121**, 443-485 (2006), http://www.santilli-foundation.org/docs/Lie-admiss-NCB-I.pdf
[24] Ktorides, C. N., Myung, H. C., and Santilli R. M.: Elaboration of the recently proposed test of Pauli's principle under strong interactions, Physical Review D **22**, 892-907 (1980), https://www.santilli-foundation.org/docs/santilli-527.pdf
[25] Santilli, R. M.: Insufficiencies of the 20th-century theories and their negative environmental implications, arXiv-0611253 (2006), https://arxiv.org/pdf/physics/0611253.pdf
[26] Santilli, R. M.: Axiomatic inconsistencies of grand unifications and their possible isotopic resolution, Annales de la Fondation Louis de

Broglie, **29**, 953-967 (2004), http://www.santilli-foundation.org/docs/inconsistencies.pdf

[27] Beghella-Bartoli, S., Muktibodh, A., and Santilli R. M.: Problematic aspects of generalized quantum theories and their apparent resolution via hadronic mechanics, Ratio Mathematica, **52,** 21-42 (2024), https://www.santilli-foundation.org/docs/2-RM-SI-Vol-52-2024-fin.pdf

[28] Biedenharn, L. C.: Angular momentum in quantum physics, Cambridge University Press (2009)

CHAPTER 6

PROGRESSIVE RECOVERING OF EINSTEIN'S DETERMINISM UNDER STRONG INTERACTIONS

6.1. Introduction

As indicated in Chapter 1, despite one century of research under large public funds, nuclear physics has been unable to achieve controlled nuclear fusion, the recycling of radioactive nuclear waste, the exact representations of nuclear data, the synthesis of the neutron from the Hydrogen atom in the core of stars; the nuclear stability despite the natural instability of the neutron; the nuclear stability despite extremely repulsive protonic Coulomb forces; and other open problems. These insufficiencies were the main motivation for the conception and construction of hadronic mechanics.

The main origin of the insufficiencies was identified in the theoretical assumption without any direct experimental verification that Heisenberg's uncertainty principle for point-like atomic constituents under sole electromagnetic interactions is also valid for extended protons and neutrons under strong nuclear interactions.

As an illustration, Heisenberg's uncertainty principle prohibits a quantitative structural representation of the synthesis of the neutron from the electron and the proton in the core of stars despite their extremely strong Coulomb attraction because the standard deviation Δr for the coordinate of the electron is much bigger than the size of the neutron and the standard deviation Δp of the electron momentum implies a kinetic energy bigger than the rest energy of the neutron. Similar structural inconsistencies are caused by Heisenberg's uncertainty principle for the solution of the remaining open nuclear problems, as we shall see in subsequent chapters.

As indicated in Sect. 3.8, the experimental evidence on strong interactions compressing nuclear constituents inside each other implies that the standard deviations for protons and neutrons in a nuclear structure must be smaller than their atomic counterparts, thus setting up the foundations for the completion of Heisenberg's uncertainty principle into a form permitting the resolution of the indicated problems of direct environmental relevance (Fig. 6.1).

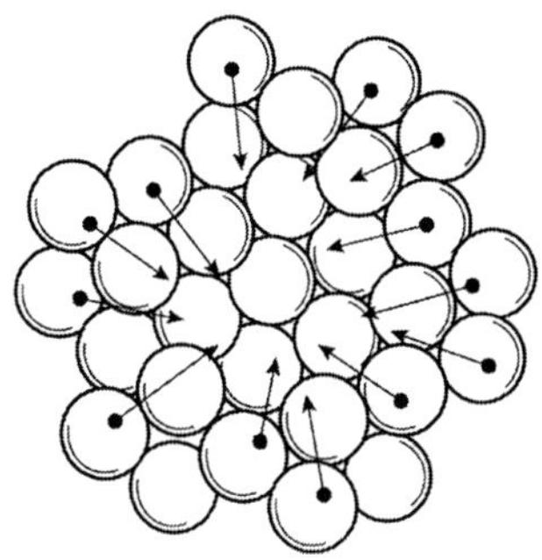

Figure 6.1. In August 1980, Santilli reported to the DOE, for referral to the White House of the nuclear physicist President Jimmy Carter, phenomenological evidence according to which the strength of strong nuclear forces implies a weakening under nuclear conditions of Heisenberg's uncertainties for atomic electrons, with potential implications for controlled nuclear fusions (see Refs. [3-22] [3-27] and Fig. 3.5). In this chapter, by following independent reviews [6-1] [6-2], we outline the laborious process over four decades that, thanks to the prior construction of hadronic mechanics [4-4] [4-5] [4-6], lead to the rigorous formulation of the uncertainties under strong interactions which is known, in its simplest possible form, as as Einstein's isodeterminism.

By following the conceptual review [6-1] and the technical review [6-2], in this chapter, we outline the laborious process according to which the achievement of maturity in isomathematics and isomechanics (Chapter 4), allowed the progressive recovering of Einstein's determinism under strong interactions.

To prevent insidious inconsistencies, quantities treated with conventional Hilbert spaces over conventional numeric fields are indicated with ordinary letters such as time t, relative coordinates r momenta p, accelerations a, energy E, density d, temperature τ, pressure π, wave functions ψ, etc. In contrast, the corresponding quantities represented on the iso-Hilbert isospace over an isofield of isonumbers are indicated with an upper star, such as t^*, r^*, p^*, a^*, E^*, d^*, τ^*, π^*, ψ^*, etc. Our studies are hereon based on the following:

6.1.1. The realization of the universal isoproduct A ⋆ B = ASB with the Santillian S = Diag. (a, b, c, d exp.{−Γ}>0 with corresponding isounit $1^* = 1/S^*$ where the quantity Γ is positive-definite but otherwise possesses an unrestricted functional dependence on ordinary variables such as Γ (t, r, p, a, E, d, τ, π, ψ, …).

6.1.2. The infinite class of isounitarily equivalent hadronic theories with time evolution U^* (t) verifying the isounitary law U^* (t) ⋆ $U^{*\dagger}$(t) = $U^{*\dagger}$(t) ⋆ U^* (t) = 1^*.

6.1.3. The dimension and shape of the individual nucleons are represented by the semi-axes a, b, and c with normalization for the perfect sphere a = b = c = 1.

6.1.4. The density of nuclear matter is represented by the parameter d with normalization d = 1 for the vacuum, with consequential increasing density in the transition from the structure of hadrons, nuclei, stars, and black holes.

6.1.5. The nonlinear, nonlocal, and non-potential interactions between extended nucleons are represented by the quantity Γ > 0 under the conditions that the Santillian S tends to 1 for mutual distances bigger than the range 1 f m = 10^{-13} cm of the strong interactions, Lim $S_{r>>1\ fm}$ = 1 and tends to 0 for infinite density, such as that occurring in black holes, Lim $S_{d\rightarrow\infty}$ = 0.

6.2. Representation of Bohm's hidden variables

In support of Einstein's view on the ultimate deterministic structure of the universe [1-1], D. Bohm [6-3] submitted in 1952 the hypothesis that certain non-local quantities called 'hidden variables' are contained in the structure of quantum mechanics resulting in a non-local theory [6-4] which achieves compatibility between the probabilistic nature of quantum mechanics and Einstein's determinism. However, many rigorously proved no-existence theorems established the im- possibility for Bohm's hidden variables to exist within the Copenhagen interpretation of quantum mechanics.

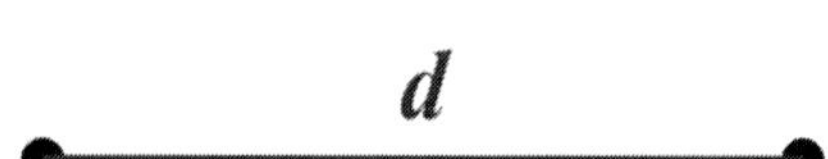

Figure 6.2. One of the most mysterious events in nature is the entanglement of particles according to which two particles initially bonded together (such as two valence electrons) and then separated, influence each other continuously, instantaneously and at arbitrary distances d without any known or visible interaction.

As indicated in Chapter 3, Bohm's hidden variables [6-2] and the de Broglie-Bohm non-local completion of quantum mechanics [6-4] [6-5] have been assumed at the foundation of hadronic mechanics thanks to the following structural steps first identified in the 1995 monographs [4-4] [4-5]:

STEP 6.2.1: By recalling that the time evolution U(t) of quantum mechanics verifies the unitary law $U U^\dagger = U^\dagger U = 1$, the sole possibility to bypass the no-existence theorems for the hidden variable is that the time evolution $U^*(t)$ of the deterministic completion of quantum mechanics must be non-unitary for the mathematics of quantum mechanics, i.e., $UU^\dagger \neq 1$, but isounitary for the isomathematics of hadronic mechanics $U = U^* S^{*,1/2}$, $UU^\dagger = U^* \star U^{*\dagger} = U^{*\dagger} \star U^* = 1^*$.

STEP 6.2.2: For Bohm's variables to be 'hidden,' it is necessary that the selected completion of quantum mechanics be based on the abstract axioms of quantum mechanics only subjected to a broader realization. The central assumption of hadronic mechanics is, therefore, the selection of its universal enveloping algebra of operators A, B, C, ... with isoproduct $A \star B = ASB$ and related isounit $1^* = 1/S$, because the resulting algebra is isoassociative, namely, it verifies the isoassociative law $A \star (B \star C) = (A \star B) \star C$ that coincides with the conventional associative law $A(BC) = (AB)C$ at the abstract, realization-free level. Consequently, the Santillian S provides an explicit and concrete realization of Bohm's hidden variables with explicit and concrete realizations of the type given Sect. 6-1.

STEP 6.2.3: For hidden variables to have physical value, they must allow numerically exact and time-invariant representations of experimental nuclear data, which representations have been impossible to date for quantum mechanics, as shown to be the case in Chapter 8.

Based on the above assumptions, Bohm's hidden variable λ can be explicitly identified with the Santillian S in all quantum mechanical associative products AB of via the application of the isounitary transformation $U^* * U^{*\dagger} = 1/\lambda = 1/S > 0$ to all quantum mechanical products $U^* \star (AB) \star U^{*\dagger} = A^* \star (U^* \star U^{*\dagger})^{-1} * B^* = A^* * S * B^*, = A^* * \lambda * B^*$, $A^* = U^* \star A \star U^{*\dagger}$, $B^* = U^* \star B \star U^{*\dagger}$ by therefore confirming the result of monographs [4-4] [4-5] according to which the Santillian is an explicit and concrete realization of Bohm's hidden variables.

6.3. Hadronic entanglements

According to experimental evidence well known since Einstein's times, two particles initially bounded together and then separated can influence each other continuously and instantaneously at arbitrary distances [6-6] via an invisible, yet real and concrete interaction, called the *"spooky action"* by Einstein. Consider for instance two electrons initially bonded into a valence pair and then separated at large mutual distance, e.g., one electron is kept in laboratory and the second electron is moved to the opposite side of Earth. In the event the laboratory electron is subjected to a spin flip, the second electron instantly experiences the same spin flip without any visible interaction, an event that Einstein called *"The spooky action at a distance"* (Fig. 6.2).

According to recent experimental evidence, said entanglement also occurs at the classical level [6-21], by therefore allowing a quantitative representation of events, e.g., in medicine [3-29], that were generally considered to be supernatural.

Albert Einstein strongly objected to the term" quantum entanglement" because the sole possible representation of particle entanglements via the Copenhagen interpretation of quantum mechanics would require superluminal communications that violate special relativity.

To honor the generally forgotten Einstein's view, Santilli [6-7] proved that the possible representation of particle entanglements by the Copenhagen interpretation of quantum mechanics is that for which the particles are free. In fact, the fundamental Schroedinger equation of quantum mechanics for a bound state of two particles with masses m_1, m_2 at mutual distance r (for $\hbar = 1$) is given by $[\Sigma_{k=1,2}\, 1/2m_k\, p_k\, p_k + V(r)]|\psi\rangle = \Sigma_{k=1,2}\, 1/2m_k\, p_k\, p_k|\psi\rangle = E|\psi\rangle$ since the potential V is identically null for all known interactions of point-like particles (such as the electrons) at large mutual distances. Consequently, the sole possible quantum mechanical representation of entanglements is that the particles are free.

By recalling that the wave packet of particles is identically null solely at infinite distance, Ref. [6-7] pointed out that the sole interactions that are continuous, instantaneous, and at arbitrary distances are given by the mutual penetration of wave packets of particles which, being non-linear, non-local and non-potential, are beyond any hope of treatment via quantum mechanics.

Thanks to the prior development of isomathematics [4-2], Santilli [6-7] proposed the axiom- preserving completion of quantum entanglements into hadronic entanglements under the suggested name of EPR entanglements, which do indeed represent particle entanglements with non-zero, yet non-Hamiltonian interactions. In fact, by recalling the realization of the isolinear isomomentum for hadronic mechanics (Sect. 4.5) $p* \star |\psi*\rangle = -i\partial r*|\psi*\rangle = -i1*\partial r|\psi*\rangle$, the hadronic equation for two entangled particles shows the clear emergence of non-linear and non-local interactions despite the identically null value of all potentials $\Sigma k=1,2\ 1/2mK\ p*k \star p*k \star |\psi 1* \star \psi 2*\rangle = [-\Sigma k=1,2\ \ 1*2/2ml\ \partial r\partial r - \Sigma k=1,2\ 1*/2mk(\partial r1*)\partial r]|\psi 1* \star \psi 2*\rangle = E\ |\psi 1* \star \psi 2*\rangle$, while verifying the superposition principle on isospace, resulting in a new interaction between particles for the possible characterization of the so-called 'fifth' interaction (Fig. 6.3).'

It should be indicated that hadronic entanglements can be uniquely and unambiguously characterized via the isounitary transformation $U*U*\dagger = U*\dagger U* = 1/\lambda = 1/S$ of the quantum entanglement $U* \star (H|\psi\rangle) \star U*\dagger = H* \star |\psi*\rangle = U* \star (E|\psi\rangle) \star U*\dagger = E* \star |\psi*\rangle$.

Note also that the completion of the quantum mechanical wave function $|\psi - 1\psi 2\ \rangle$ into the broader wave function of hadronic mechanics $|\psi 1* \star \psi 2*\rangle$ confirms Einstein's view that "The wave function of quantum mechanics does not provide a complete description of the entire physical reality" [1-1].

6.4. Generalization of Bell's inequalities for strong interactions

In 1964, J. S. Bell [6-8] published a theorem according to which, under the assumption of various quantum axioms, systems of point-like particles with spin 1/2 do not admit a classical counterpart by preventing the existence of hidden variables and Einstein determinism. The theorem was proved by showing that a certain expression D^{Bell} (whose explicit value depends on the relative conditions of the particles) is always smaller than the corresponding classical value D^{Clas}.

By representing the extended character of nuclear constituents via the isotopic completion of Pauli's matrices with Bohm's hidden variables (Eqs.

(6.8.20), p. 254 of [4-5]), today called the Pauli-Santilli isomatrices, Santilli [6-9] proved in 1998 that hadronic mechanics implies the generalization of Bell's inequalities for strong interactions into a form D^{HM} which always admit a classical counterpart, $D^{HM} = D^{Clas}$, by therefore confirming Einstein's view on the possible recovering of classical determinism.

In essence, the above studies established that Bell's inequalities are indeed valid for electromagnetic interactions of point-like particles, with an ensuing lack of hidden variables and a lack of exact representation of nuclear data.

However, Bell's inequalities are inapplicable in favor of their completions by hadronic mechanics for systems of extended particles at short mutual distances. The exact representation of their experimental data requires a non-unitary completion of quantum models, under which none of Bell's assumptions can be formulated.

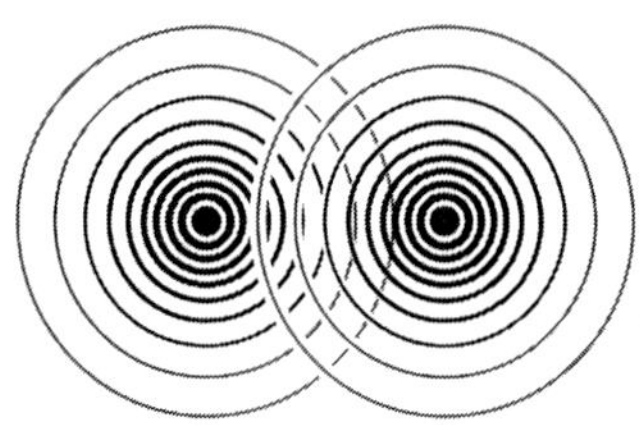

Figure 6.3. Santilli [6-7] did achieve the first known quantitative representation of particle entanglements by admitting the only possible interaction between entangled particles, that caused by the overlapping of the wave packets of particles which interaction, however, is non-linear, non-local and non-potential, thus explaining the prior need to develop isomathematics [4-2] and the decades of studies for the achievement of the indicated result.

It then follows that experiments [6-10] [6-11] [6-12] proving the violation of Bell's inequalities are direct experimental verifications of hadronic mechanics as shown in the recent paper [6-14].

Note that, as was the case for hidden variables and hadronic entanglements, the hadronic quantity D^{HM} can be uniquely and unambiguously characterized by the step-by-step isounitary transformation $U^* \star U^{*\dagger} = 1/\lambda$

= 1/S of the entire procedure yielding D^{Bell}, here symbolically written $D^{HM} = U^* \star D^{Bell} \star U^{*\dagger}$.

6.5. Einstein's isodeterminism

Soon after joining Harvard University in late 1977, Santilli doubts the validity of strong interactions between Heisenberg's uncertainty principle and other quantum mechanical laws, as one can see from the title of the 1978 memoir [3-5]. Santilli's argument underlying such a view is that Heisenberg's standard deviations for coordinates Δr and momenta Δp are valid for the conditions of their original conception, i.e., for point-like particles in vacuum under electromagnetic interactions, since a point-like particle can move within a star by solely sensing action-at- a-distance interactions due to its dimensionless character.

The situation is mathematically, theoretically, and experimentally different when considering extended nucleons in conditions of mutual penetration because, given their 'strength', strong interactions imply the creation of pressure on a given extended nucleon caused by its surrounding nucleons that does not exist in an atomic structure (Fig. 6-4).

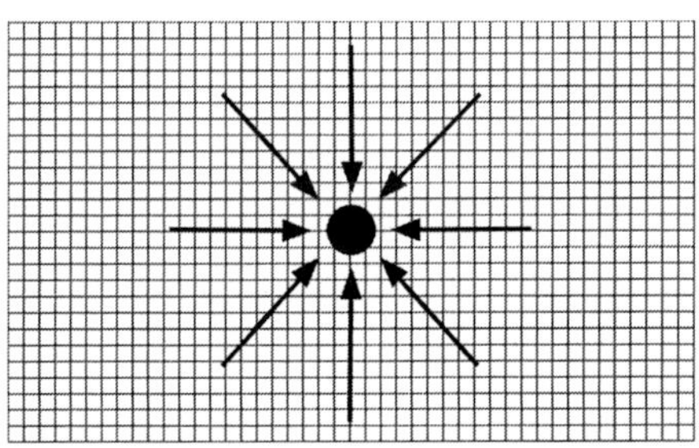

Figure 6.4. The study of nuclear structures as existing in nature, i.e., composed by extended nucleons in conditions of partial mutual penetration, implies the emergence of new physical quantities that have to be represented which are simply inconceivable for quantum mechanics, such as the pressure exercised on an individual nucleon by all remaining nucleons, which pressure is ultimately implicit in the non-local conception by the Duke Louis de Broglie that the characteristics of a particle should depend on the action by all surrounding particles.

Thanks to the 1978 conception and development of the isoassociative enveloping algebra [3-5], Santilli submitted in 1981 [3-21] the hypothesis that the standard deviations for the indicated nucleons $\Delta r*$ and $\Delta p*$ have to be smaller than the corresponding deviations for an electron in a vacuum. Additionally, Santilli proved in 1994 [6-14] that Einstein's determinism [1-1] is achieved in full in the interior of gravitational collapse, by therefore establishing the decrease in value of the Santillian with the increase of the density of the hadronic medium, and its null value for gravitational collapse.

The smooth connection between the weakening of the standard deviations within nuclear matter [3-21] and their null value for gravitational collapse [6-14] was achieved in Ref. [6-15], with extensive treatments in Refs. [5-16]-[6-18] and can be summarized as follows. Assume the isonormalization of the isostates of the iso-Hilbert isospace $\langle\psi*|\star|\psi*\rangle = \langle\psi*|S|\psi*\rangle = S$, and the isolinear isomomentum $p* \star |\psi*\rangle = -\partial r*|\psi*\rangle = -1*\partial r|\psi*$, with ensuing isocommutation rules $[ri*,p*j]* \star|\psi*\rangle = -i\delta ij1*$, $[ri*,rj]* \star|\psi*\rangle = [p*i,p*j]*|\psi*\rangle = 0$.

Then, as was the case for Bohm's hidden variables (Sect. 6.2), quantum entanglements (Sect. 6.3), and Bell's inequalities (Sect. 6.4), the isounitary transformation of Heisenberg's uncertainty principle uniquely and unambiguously yields Santilli's isouncertainty principle of hadronic mechanics [6-15] $\Delta r\Delta p = (1/2)|\langle\psi*| \star [r*, p*]* \star |\psi*\rangle| = (1/2)\ S$.

LEMMA 6-5.1 (ISODETERMINISTIC PRINCIPLE [6-15]): The isostandard isodeviations for isocoordinates Δr^* and isomomenta $\Delta p*$, as well as their product, progressively approach classical determinism for extended particles in the interior of hadrons, nuclei, and stars, and achieve classical determinism at the extreme densities in the interior of gravitational collapse.

In particular, the standard isodeviations $\Delta r*$ and $\Delta p*$ progressively and individually tend to zero with the increase of the density of the hadronic medium, thus in the transition from hadrons to nuclei and stars [6-17].

To represent the full validity of Einstein's determinism, Santilli [6-19] [6-20] decomposes the Riemannian metric g(x) in four dimensions into the product of the Minkowskian metric $\eta = -\mathrm{Diag}, (1, 1, 1, -1)$ and the gravitational Santillian $g(x) = S_{4\times4}\ \eta_{4\times4}$ here assumed as the Santillian of relativistic hadronic mechanics (Chapter 4), with particular values for the Schwartzschild metric $S_{kk} = 1/(1 - 2M/r)$, $S_{44} = 1 - 2M/r$. It is then easy to see that Einstein's determinism [1-1] is fully recovered at the limit of the Schwartzschild horizon.

A yet unexplored consequence of Einstein's isodeterminism, which is important for particle physics, is the following:

COROLLARY 6.5.1 [6-16] [6-17]. Einstein's isodeterminism, according to Lemma 6.5.1, implies the removal of quantum mechanical divergencies.

To prove this corollary, it is sufficient to recall that the Santillian is inserted in all quantum mechanical products. But its value is much smaller than one. Consequently, divergent quantum mechanical perturbative series are turned into convergent hadronic forms [3-4].

References of Chapter 6

[1] Editorial Team, The progressive recovering of Einstein's determinism under strong interactions, Scientia, 2024, Link to be provided

[2] Muktibodh, A: Santilli's recovering of Einstein's determinism, *Progress in Physics,* 20, 26-34 (2024), http://www.eprdebates.org/docs/muktibodh-2024.pdf

[3] Bohm, D. A Suggested Interpretation of the Quantum Theory in Terms of 'Hid- den Variables,' *Phys. Rev.* 85, 166-182 (1952), journals.aps.org/pr/abstract/10.1103/PhysRev.85.166

[4] Stanford Encyclopedia of Philosophy, Bohmian (de Broglie-Bohm) Mechanics (2021), https://plato.stanford.edu/entries/qm-bohm

[5] Bohm, D. Quantum Theory, Prentice-Hall, New York (1951).

[6] Bengtsson I. and Zyczkowski K., *An Introduction to Quantum Entanglement,* Cam- bridge University Press, second, revised edition (2006).

[7] Santilli, R. M. A quantitative representation of particle entanglements via Bohm's hidden variables according to hadronic mechanics, *Progress in Physics* 18, 131-137 (2022), www.santilli-foundation.org/docs/pip-entanglement-2022.pdf

[8] Bell, J. S. On the Einstein Podolsky Rosen paradox *Physics* 1, 195 (1964), https://cds.cern.ch/record/111654/files/vol1p195-200 001.pdf

[9] Santilli, R. M. Isorepresentation of the Lie-isotopic SU(2) Algebra with Application to Nuclear Physics and Local Realism, *Acta Applicandae Mathematicae* 50, 177-190 (1998), www.santilli-foundation.org/docs/Santilli-27.pdf

[10] Fadel M. et al., Spatial entanglement patterns and Einstein-Podolsky-Rosen steering in Bose-Einstein condensates. *Science* 360, 409-415 (2018), www.santilli-foundation.org/Basel- paper.pdf

[11] Colciaghi P. et al., Einstein-Podolsky-Rosen Experiment with Two Bose-Einstein
Condensates, Phys. Rev. X 13, 021031-1/021031-10 (2023), /http://journals.aps.org/prx/pdf/10.1103/

[12] Aspect A. et al., Experimental Realization of Einstein-Podolsky-Rosen-Bohm Gedankenexperiment: A New Violation of Bell's Inequalities, *Phys. Rev. Lett.* 49, 91-94 (1982), //ui.adsabs.harvard.edu/abs/1982PhRvL..49...91A

[13] Beghella-Bartoli, S., A. Muktibodh, A. and Santilli R. M., Problematic aspects of generalized quantum theories and their apparent resolution via hadronic mechanics, *Ratio Mathematica*, 52, 21-42 (2024), https://www.santilli-foundation.org/docs/2-RM-SI-Vol-52-2024-fin.pdf

[14] Santilli, R. M., Isotopic lifting of Heisenberg's uncertainties for gravitational singularities *Springer Nature Communications in Theoretical physics* 3, 47-66 (1994),
http://www.santilli-foundation.org/docs/santilli-511.pdf

[15] Santilli R. M., Studies on the classical determinism predicted by A. Einstein, B. Podolsky and N. Rosen, *Ratio Mathematica*37, 5-23 (2019), http://www.eprdebates.org/docs/epr- paper-ii.pdf

[16] Santilli, R. M., Studies on A. Einstein, B. Podolsky and N. Rosen prediction that quantum mechanics is not a complete theory, Paper I: Basic methods, *Ratio Mathematica* 38, 5-69 (2020),
http://eprdebates.org/docs/epr-review-i.pdf

[17] Santilli, R. M., Studies on A. Einstein, B. Podolsky and N. Rosen prediction that quantum mechanics is not a complete theory, Paper II: Apparent proof of the EPR argument, *Ratio Mathematica* 38, 71-138 (2020), http://eprdebates.org/docs/epr-review-ii.pdf

[18] Santilli, R. M., Studies on A. Einstein, B. Podolsky and N. Rosen prediction that quantum mechanics is not a complete theory, Paper III: Illustrative examples and applications, *Ratio Mathematica* 38, 139-222 (2020), http://eprdebates.org/docs/epr-review-iii.pdf

[19] Santilli, R. M., Isotopic quantization of gravity and its universal isopoincaré symmetry, in the Proceedings of The Seventh Marcel Grossmann Meeting on Gravitation, SLAC, 1992, Jantzen, R. T., Keiser, G. M. and Ruffini, R., Editors, *World Scientific Publishers,* 500-505 (1994),
www.santilli-foundation.org/docs/Santilli-120.pdf

[20] Santilli, R. M., Unification of gravitation and electroweak interactions, in the Proceedings of the Eight Marcel Grossmann Meeting on

Gravitation, Israel, 1997, T. Piran, and Ruffini, R. Editors, World Scientific, p. 473-475 (1999), www.santilli-foundation.org/docs/Santilli- 137.pdf

[21] Berkowitz, R.: Macroscopic systems can be controllably entangled and limitlessly measured, *Physics Today*, July issue, p. 16-18 (2021), www.santilli-foundation.org/docs/Berkowitz-paper.pd

CHAPTER 7

HADRONIC SYMMETRIES AND RELATIVITIES

7.1. Introduction

Galileo's and Einstein's special relativities are hereon assumed to be fully valid under the conditions of their original conception, namely, for point-like approximation of particles and their wave packets while moving in vacuum under sole action-at-a-distance, thus potential interactions. In this chapter, we hope to indicate that more general realizations of the abstract axioms of said relativities may significantly broaden their applicability to extended particles within physical media as well as to open problems in gravitation.

To begin our study, let us recall, that the fundamental notion underlying Galileo and Einstein's special relativity is the Newtonian notion of massive point in vacuum due to the local character of the Newton-Leibnitz differential calculus, that is, its sole possible definition at isolated points in space.

But massive points cannot experience any possible resistive force. Consequentially, the fundamental equations of Galileo's and Einstein's relativities and their space-time symmetries is Newton's equation restricted to potential, thus variationally selfadjoint forces [3-8] $m(dv/dt) - F^{SA}(r) = 0$ on the Euclidean space $E(r., \delta, 1)$ with multiplicative unit 1 and metric $\delta =$ Diag. $(1, 1, 1)$ over the field of real numbers R.

Despite its historical character, the reduction of the entire universe to point-like particles in a vacuum are easily dismissed by a mere inspection of nature, let alone a significant number of generally ignored experimental dismissals reviewed in Chapter1 since particles may have a point-like charge but point-like wave packets do not exist in nature.

This evidence establishes the need to identify broader symmetries and relativities for characterizing extended particles/wave packets in conditions of mutual overlapping and entanglement with ensuing linear and non-linear, local and non-local, and potential and non-potential interactions whose study was initiated by Santilli in the late 1970's at Harvard University under DOE support (see Chapter 3 and Sect. 3.7 in particular). Said need is established in various fields, e.g., in nuclear physics, where measured nuclear volumes are generally smaller than the sum of the measured volumes of the nucleon charge distributions, resulting in an inevitable non-Hamiltonian component of the nuclear force. (Fig. 1.3)

The suggested study is also necessary for the credibility of the experimental verifications of any completion of quantum mechanics, which credibility, as indicated in preceding chapters, depends on broader theories being able to predict the same numeric values under the same conditions at all times, a feature generally referred to as invariance which can only occur under the identification of the new symmetries for Hamiltonian and Santillian interactions.

In view of the above evidence, the main aim of hadronic mechanics is the time invariant representation of "extended" particles, (such as a balloon at the classical level or a nucleon at the operator level), thus requiring a representation of their dimension, shape and density, while moving within a hadronic medium, (e.g., air at the classical level or core of a star at the operator level).

As it is well known, the full historical Newton's equations do admit contact non-potential (e.g., velocity-dependent), thus variationally non-selfadjoint forces, $m(dv/dt) - F^{SA}(r) - F^{NSA}(t, r\ v,...) = 0$ [3-8], but the underlying differential calculus for the crucial characterization of the acceleration $a = dv/dt$ remains solely valid for massive points.

Following the achievement of maturity in the construction of isomathematics, Santilli completed Newton's equations in memoir [4-2] of 1996 into the Newton-Santilli isoequations here formulated with respect to absolute time t, $m^* \star (d^*v^*/d^*t) - F^{*SA}(r) - F^{*NSA}(t, r^*, vv, ...) = 0$, where $m^* = m\ 1^*$ represents the extended mass, which equation is defined on an iso-Euclidean isospace $E^*(v^*, \delta^*, 1^*)$ with isounit representing dimension, shape and density of particles (Chapter 6) $1^* = Diag.(a, b, c, d)\ \Gamma = 1/S < 0$, isometric $\delta^* = S\ \delta$ and isovelocities $v^* = v\ 1^*$ over an isofield $F^*(n^*,^*, 1^*)$ of real isonunbers $n^* = mn\ 1^*$ with isoprodct $n^* \star k^* = n\ k\ 1^*$, the Newtonian forces are now isofunctions, and Newton's velocity-dependent forces are now represented with the isodifferential $d^*v^* = dv - v\ dS$.

The complexity of the problem here considered is that, while quantum mechanics solely admits local symmetries and relativities, hadronic mechanics admits the following chain of mathematically and physically inequivalent hadronic symmetries and relativities:

7.1.1. Local symmetries and relativities, Galileo's and Einstein's special relativity.

7.1.2. Lie-isotopic relativities, or Isorelativities for short, which are characterized by the Lie-isotopic mathematics and are intended to describe closed, thus conservative and time reversal invariant systems of extended particles in conditions of mutual penetration with Hamiltonian and Santillian interactions under the condition of admitting local relativities as particular cases.

7.1.3. Lie-admissible symmetries and relativities, or genorelativities for short, which are characterized by the Lie-admissible mathematics and are intended to describe open, thus non-conservative and time irreversible systems of extended particles with Hamiltonian and Santillian interactions under the condition of admitting the preceding relativities as particular cases.

7.1.4. Hyperstructural symmetries and relativities, or Hyperrelativities for short, which are characterized by the multi-valued Vougiouklis Lie-admissible mathematics for the description of biological, thus irreversible systems while admitting all preceding relativities when their multi-valued hyperunit is reduced to single values.

7.1.5. Isodual symmetries and relativities or Isodual local, iso-, geno- and hyper-relativities which are characterized by the isodual mathematics with negative units and intended to describe antimatter in various conditions.

Note that the above relativities and related symmetries should be formulated at the classical and operator levels and for non-relativistic and relativistic conditions.

7.2. Rudiments of the Lie-Santilli isotheory

Recall that an N-dimensional Lie symmetry G is characterized by Hermitean generators X_k, $k = 1, 2, ..., N$ verifying the Lie algebra commutation rules $[X_i, X_j] = X_i \times X_j - X_j \times X = C^k_{ij} \times X_k$ over a numeric field

$F(n, \times, 1)$. Recall that the infinite family of isotopes G^* of G for all possible values of S admits the isoalgebra with isocommutation rules $[X_i, X_j]^* = X_i \star X_j - X_j \star X_i = X_i S X_j - X_j S X_i = C^k_{ij} S X_k$ over F^*, where the generators X_k are the same as those of G since they represent total, physical, measured, conserved quantities. Recall also that a simple transition from a Lie algebra to a Lie transformation group of a local coordinate $x' = \Lambda \times x$ can be done via the exponentiation $\Lambda = expX$, where the generator X is generally expressed in its matrix form.

Similarly, a simple transition from a Lie-Santilli isoalgebra to its corresponding Lie-Santilli isogroup of isotransformations of a local isocoordinate $x'^* = \Lambda^* \star x^*$ on F^* can be done via the isoexponentiation [3-9] $e^* = exp^*\{X\} = [exp\{XS\}] \times 1^* = 1^* \times [exp\{SX\}]$ [4-4] where X is the generator of G and S represents the selected non-potential interactions, thus being fixed for a given system as it is the case of the Hamiltonian.

The explicit construction of Lie-isotopic symmetries is then completed by the following property known as the fundamental theorem of isosymmetries [3-9] [4-12] [7-1]:

THEOREM 7.2.1: Let G be an N-dimensional Lie symmetry on a K-dimensional space E(x, m, F) with local coordinates x, metric m, and parameter w over a numeric field $F(n, \times 1)$ with transformations $x' = A(w) \times x$, $y' = A(w) \times y$, $x, y \in E$, leaving invariant the line element $(x - y)^t \times m \times (x - y) \equiv (x' - y')^t \times A^t(w) \times m \times A(w) \times (x' - y')$ or, equivalently $A^t(w) \times m \times A(w) \equiv m$. Then, all infinitely possible isotopies G^* of G on isospace $E^*(x^*, m^*, F^*)$ with isocoordinates $x^* = x1^*$, isometric $m^* = S \times m$ and isotransformation $x'^* = A^*(w^*) \star x^*$, $y'^* = A^*(w^*) \star y^*$, $x^*, y^* \in E^*$ over a numeric isofield F^* with positive-definite Santillian S, leave invariant the isoline element $(x^*-y^*)^t \star m^* \star (x^*-y^*) \equiv (x'^*-y'^*)^t \star A^{*t}(w^*) \star m^* A^*(w^*) \star (x'^*-y'^*)$ or, equivalently, $A^{*t} \star m^* \star A^* \equiv m^*$, and all infinitely possible so constructed Lie-Santilli isosymmetries G^* are locally isomorphic to the original symmetry G.

Irreversible Lie-admissible symmetries and relativities formulated on a genospace ${}^{<}U \times U^{>}$ with corresponding, time-dependent, invertible Santillians ${}^{<}S$ and $S^{>}$ can be studied with the same Theorem 7.1 restricted to forward motion with its anti-Hermitean image for backward motion along studies initiated in monographs [4-4] [4-5].

The hyperstructural symmetries and relativities are expected to follow via the hyperstructural broadening of the Lie-admissible mathematics along studies initiated by T. Vougiouklis [4-3]. Isodual symmetries and relativities are expected to follow studies initiated in the monograph [4-11].

Note that broadening the above studies via "non symmetric" metrics [3-6] [3- 7] is particularly intriguing since they would geometrize irreversibility via the metrics themselves and induce new geometries.

Needless to say, we cannot possibly review these studies in detail and can only provide a guide to the existing literature and a glimpse of possible future developments in the field.

7.3. Santilli's Galilean isorelativity

As set in history beginning with tests at the Pisa tower, the Galilean symmetry and relativity provide a time invariant description of massive points moving in vacuum with constant speed v under sole potential interactions, by therefore leaving invariant Newton's equation $m(dv/dt) - F^{SA}(r) = 0$ and basic transformation $r' = r - vt$.

The Galilean isorelativity (Fig. 2.4) is intended to provide an invariant description over absolute time t of the motion of extended masses m* within a medium characterized by the Santillian S with ensuing potential and non-potential interactions, by therefore leaving invariant the Newton-Santilli isoequation $m^* \star (d^*v^*/d^*t) - F^{*SA}(r) - F^{*NSA}(t, r^*, v^*, ...) = 0$ with basic isotransformation $r'^* = r^* - v^* \star t$ (see the first classical proposal in Chapter 6 of the 1983 monograph [3-9], the extension to operator formulations in the 1991 monographs [7-2] [7-3]; the notes of Santilli's 1991 lectures at the ICTP, Trieste Italy [4- 19]; the Nobel Nomination [7-4]; and successfully applied to various non-relativistic problems addressed by hadronic mechanics and progressively reviewed in this book..

Figure 7.1. A light beam passing through water is generally reduced to photons propagating in vacuum to maintain special relativity in interior conditions. However, such a reduction is afflicted by the generally ignored Inconsistencies 1) to 7) of Section 7-4-1.

The Galilean isorelativity was constructed in monograph [3-9] according to the following main steps: 1) Construction of the isotopies of the various branches of Lie's theory (Lie-Santilli isotheory); 2) Construction of Birkhoffian mechanics with Hamiltonian mechanics as a particular case; 3) Proof that Birkhoff's equations are universal for the representation of all possible (regular) NSA Newtonian systems; 4) Show that the generalized brackets of the time evolution of Birkhoff's equation provide a realization of the Lie-Santilli isotheory; 5) Construction of the classical realization of the iso-Galilean symmetry $G^*(3.1)$ and relativity via Theorem 7.1 in terms of the Lie-isotopic brackets of Birkhoff's equations

Nowadays, isomathematics allows a rather simple construction of the Galilean isosymmetry and relativity. Assume the simple realization of the Santillian $S = exp\ \{-\Gamma(t,\ r,\ v)\}$. Then, the Newton-Santilli isoequation becomes $m^* \star (d^*v^*/d^*t) - F^{SA} = 1^*\ [m(dv/dt) + mvS(d1^*/dt\ ) - F^{SA}] = 1^*\ [m(dv/dt\ ) - F^{SA} - F^{NSA} = 0]$ with representation of the NSA forces by the Γ function

$mvS(d1^*/dt) = mv(d\Gamma/dt) = F^{NSA}$ from which $\Gamma = -(1/mv) \int F^{NSA}dt$ and $S = \exp\{-(1/mv) \int F^{NSA}dt\}$ [4-2]. The knowledge of the Santillian then allows the construction of the isotope $G^*(3.1)$ of the Galilean symmetry $G(3.1)$ [7-2] with the invariance of the NSA Newtonian system assumed by Theorem 7.2.1. The explicit construction of the iso-Galilean isotransformations can then be done via the simple method outlined in Sect.1.

The most important application of the Galilean isosymmetry and isorelativity is their crucial role in achieving the first known numerically exact and time-invariant non-relativistic representation of nuclear data reviewed in Chapter 8. At this point, we merely mention that contrary to a rather popular belief, non-relativistic treatments are effective for nuclear structures because the strength of strong nuclear forces severely restricts the dynamics of internal protons and neutrons, as correctly represented by the hadronic uncertainty principle and the small numeric value of the Santillian $S = exp\ \{-\Gamma\} \ll 1$ studied in the preceding chapter.

7.4. Santilli's special isorelativity

7.4-1. Foreword

In preparation for the construction of the relativities for extended particles propagating within physical media, the author constructed the isotopies of all continuous and discrete space-time symmetries, including in chronological order:

1983, isotopies of the Lorentz symmetry [7-5];
1983, isotopies of operator symmetries [7-6];
1985, isotopies of the rotational symmetry [7-7] [7-8];
1993, isotopies of the SU(2) symmetry [7-9];
1991, isotopies of the Poincare symmetry [7-10][7-11];
1995, isotopies of the spinorial covering of the Poincare symmetry [7-12];
1998, isotopies of the Minkowskian geometry.

Lie-isotopic symmetries, also called isosymmetries [7-5]-[7-13], have been applied by the author to several open problems in physics outlined in this book and independently studied by numerous scholars (see, e.g., monographs [4-12]-[4-21] and references quoted therein).

In this section, we provide an elementary review of the most important space-time Lie isotopic symmetries and related relativities called isorelativities with the understanding that their technical knowledge can be solely acquired by inspecting their originating papers.

Regarding terminologies, we shall use the word ″isosymmetries″ when the new invariances are formulated on isospaces over numeric isofields and call them ″symmetries″ when they are projected into our space-time over a conventional numeric field.

7.4-2 Inapplicability of special relativities within physical media

It has been widely believed for one century that special relativity and the constancy of the speed of light *c* are valid for whatever conditions exist in the universe while ignoring Einstein's repeated statement on the constancy of the speed of light *c* ″in space,″ thus implying the distinction between ″exterior dynamical problems″ of point particles in vacuum and the ″interior dynamical problems″ for extended particles within physical media, which distinction has been widely ignored by mainstream academia for the past century.

As indicated earlier, in these studies, we assume the validity of space-time symmetries and relativities for the exterior dynamical problems and study their completion or extension for interior problems. This study is motivated by serious insufficiencies or sheer inconsistencies in applying space-time Lie symmetries and relativities to interior problems that should be addressed

As an example, for the intent of maintaining the validity of special relativity in water, it is widely assumed that light propagating in water can be reduced to photons scattering among the water molecules by, therefore, propagating in a vacuum while generally ignoring several ensuing insufficiencies, such as [1-4]:

INSUFFICIENCY I: The reduction of light to photons cannot represent the angle of refraction of light in water, evidently because photons will scatter in all directions at the impact with the water surface.

INSUFFICIENCY II: The reduction of light to photons cannot provide a numerical representation of the decrease in the speed of light in water by 100 000 km/s.

INSUFFICIENCY III: Reducing a beam of light to photons cannot represent light propagation in water as a beam because photons will scatter among the water molecules and disperse in various directions.

INSUFFICIENCY IV: The reduction of light to photons has no physical meaning for infrared light or radio waves with 1 *m* wavelength that experience the same phenomenology as that for electromagnetic waves at large.

INSUFFICIENCY V: The reduction of light to photons scattering among water molecules, thus propagating in vacuum at speed *c*, cannot represent the Cherenkov light because said light can only occur when electrons travel faster than the local speed of light.

INSUFFICIENCY VI: Reducing light to photons violates the relativistic sum of speeds because the sum of two light speeds in water does not yield the light speed in water.

INSUFFICIENCY VII: Light reduction to photons cannot be tested experimentally due to the lack of an inertial reference frame in water.

Given the above insufficiencies, a primary objective of the isotopies of special relativity is the achievement of the invariant representation of extended particles and electromagnetic waves propagating within physical media.

As a historical note, it should be mentioned that Albert Einstein received his 1921 Nobel Prize in Physics for the quantized absorption of light and not for its quantized propagation.

7.4-3. Lorentzian isosymmetry

As it is well known, H. Lorentz attempted 1904 the construction of the invariance of the speed of light of his time within physical media $C = c/n$ but failed to do so and had to restrict his study to the invariance of the constant speed of light c in vacuum.

Figure 7.2: A view of the apparent increase of size of objects submerged in water as seen by an outside observer which is numerically represented by special isorelativity with intriguing geometric implications.

Lorentz's restriction to the invariance of the constant speed c was due to Lie's theory's linear, local, and Hamiltonian character. At the same time, Lorentz's problem was highly non-linear, non-local, and non-Hamiltonian due to the complex functional dependence of the index of refraction n on local variables, such as density, temperature, etc. (see, e.g., Eq. (11), p. 266 of [7-5]). For this reason, Santilli constructed in monograph [3-9] the Lie isotopic completion of Lie's theory for non-linear, non-local, and non-Hamiltonian systems, and achieved in the 1983 paper [7-5] the universal invariance SO^* (3.1) of Lorentz's speed of light $C = c/n$ via the following main steps.

Recall that the vacuum is geometrically represented in special relativity via the Minkowski space $M(x, \eta, 1)$ with coordinates $x = (x^1, x^2, x^3, x^4 = ct)$, metric $\eta = Diag.\,(1, 1, 1, -1)$ and line element $x^2 = x^\mu \times \eta_{\mu\nu} \times x^\nu$. Santilli proposed in paper [7-5] the geometrization of physical media via the infinite class of iso-Minkowski spaces $M^*(x^*, \eta^*, 1^*)$ characterized by the isounit $1^* = 1/S > 0$, space-time isocoordinates $x^* = x^* 1^*$ and isometric $\eta^* = S\eta$ with the realization of the Santillian $S = Diag.\,(1/n^2{}_1, 1/n^2{}_2, 1/n^2{}_3, 1/n^2{}_4,)$ where, in this application, $n^2{}_1$, $n^2{}_2$, $n^2{}_3$ represent the possible inhomogeneity and anisotropy of the medium and $n^2{}_4$ its density, resulting in the isoline element $x^{*2} = x^{*\mu} \bigstar \eta^*{}_{\mu\nu} \bigstar x^{*\nu} = (x^{1,2}/n^2{}_1 + x^{2,2}/n^2{}_2 + x^{3,2}/n^2{}_3 - x^{4,2}/n^2{}_4)1^*$.

The Lie-Santilli isoalgebra so^* (3.1) of the isosymmetry SO^* (3.1), today known as the Lorentz-Santilli isoalgebra and isosymmetry, is characterized by the isogenerators $J^*{}_1 = n_2 n_3 J_1$, $J^*{}_2 = n_1 n_3 J_2$, $J^*{}_3 = n_1 n_2 J_3$, $M^*{}_k = n_k M_k$, where J_k, M_k are the conventional generators of $so(3.1)$, with isocommutation rules $[J^*{}_i, J^*{}_j]_* = K^2 \epsilon_{ijk} J^*{}_k$, $[M^*{}_i, M^*{}_j]^* = \epsilon_{ijk} J^* k$, $[J^*{}_i, M^*{}_j]^* = -\epsilon_{ijk} M^*{}_k$, and iso-Casimir isoinvariant $C^*{}_1 = J^{*2} - 1/K\, M^{*2} = -3 1^*$, $C^*{}_2 = J^* \bigstar M^* = 0$.

By denoting the generators J_k, M_k, $k = 1, 2, 3$ with the unified symbol X_k, $k = 1, 2, \ldots, 6$, a simple transition from the isoalgebra so^* (3.1) to the isogroup SO^* (3.1) can be done via the isoexponentiation $\Lambda^*(w) = \{\Pi^*{}_k exp^*(iX_k w_k)$ $= \{\Pi_k exp(iX_k S w_k)$ resulting in the transformations leaving invariant the iso-Minkowskian isoline element, first introduced in [7- 5] and here reported for simplicity in the (3, 4)-plane (see Chapter 8, p.329 on of Ref. [4-5] for a general treatment) $x^{1'} = x^1$, $x^{2'} = x^2$, $x^{3'} = \gamma^* (x^3 - \beta^* n_3/n_4\, x^4)$, $x^{4'} = \gamma^* (x^4 - \beta^* n_4/n_3\, x^3)$, where $\beta^* = (v_3/n_3)/(c/n_4)$, $\gamma^* = 1/\sqrt{1 - \hat{\beta}^{*,2}}$.

By using the isodifferential calculus [4-2] and assuming for simplicity that $t^* = t$, we can write the SO^* (3.1) invariant expression $dx^{3,*}/d^*dt = C = c/n_4$) under which the 1983 paper [7-5] solved, for the first time, the Lorentz historical problem of the invariance of the local speed of light $C = c/n$, $n = n_4$.

A few comments are now in order. The indicated Lorentz-Santilli transformations had been presented in their projection into the conventional Minkowski space M over the conventional field of real numbers R because they formally coincide with the conventional Lorentz transformations when formulated on isospace M* over R*.

Despite the local variation of the speed of light within physical media and given the local isomorphism from Theorem 1.1, SO^* (3.1) ≡ SO(3.1), the Lorentz-Santilli transformations characterize a perfect light cone on M^* over R^*, namely, the maximal causal speed on M^* over R^* is c, by therefore restoring the exact Lorentz symmetry when popularly believed to be" broken." Paper [7-5] has established the universality of the Lorentz symmetry for all infinitely possible symmetric line elements with a metric in (3+1)-dimensions with an arbitrary (regular) dependence on local coordinates [7-14].

Note that the Lie-Santilli isotheory does not restrict the density of the medium besides that of being positive-definite, $n_4 > 0$. As such, n4 can be smaller or greater than one. Consequently, the Lorentz-Santilli isosymmetry characterizes causal subluminal speeds of massive extended particles/wave packets within physical media of low density, $C = c/n_4 < c$, $n_4 > 1$ (as it is the case for water) and causal superluminal speeds for massive extended particles/wave packets within physical media of high-density $C = c/n_4 > c$, $n_4 < 1$ (as predicted for the interior of hadrons nuclei and stars), whose experimental confirmation is reported in Section 7.3.9 and the subsequent Chapters.

Alternatively, from the local isomorphism SOI (3.1) ≡ SO^* 3.1), we can say that the completion of the Lorentz symmetry into the Lorentz-Santilli isosymmetry has essentially proved that the abstract axioms of special relativity do indeed admit arbitrary speeds, with a consequential considerable increase of its representational capability [7-18].

7.4-4. Poincarèian isosymmetry.

The second basic symmetry of relativistic hadronic mechanics is given by the isotopies P^* (3.1) = SO^* (3.1) $\bigstar$ T^* (3, 1) of the Lorentz-Poincarè symmetry $P(3.1) = SO(3.1) \times T(3, 1)$, where $T(3.1)$ represents the symmetry under translations, which was introduced in the 1991 paper [7-10] (see also the 1993 paper [7-11], Sect. 8.3, p. 342 on of monograph [4-5], and independent studies [4-19]), today called the Lorentz-Poincarè Santilli isosymmetry.

The isoline element left invariant by P^* (3.1) is given by $(x^* - y^*)^{2*} = (x^{*\mu} - y^{*\mu}) \bigstar \eta^*_{\mu\nu} \bigstar (x^{*\nu} - y^{*\nu}) = [(x^1 - y^1)^2/n^2_1 + (x^2 - y^2)^2/n^2_2 + (x^3 - y^3)^2/n^2_3 - (x^4 - y^4)^2/n^2_4)1^*$ $x, y \in M^*$ The isoalgebra p^* (3.1) of P^* (3.1) is characterized by the Hermitean generators ($J_{\mu\nu} = (J_k, M_k)$ and isocommutation rules $[J_{\mu\nu}, J_{\alpha\beta}]^* == \imath(\eta^*_{\nu\alpha} J_{\beta\mu} - \eta^*_{\mu\alpha} J_{\beta\nu} - \eta^*_{\nu\beta} J_{\alpha\mu} + \eta^*_{\mu\beta} J_{\alpha\nu})$, $[J_{\mu\nu}, P_\alpha]^* = i(\eta^*_{\mu\alpha} P_\nu \eta^*_{\nu\alpha} P_\mu)$, $[P_\mu\hat{}, P_\nu]^* = 0$, and iso-Casimir-isoinvariants $C^*_1 = 1^* > 0$, $C^*_2 = P^{*2} = P_\mu \bigstar P_\mu = (\eta^*_{\mu\nu} P^\mu S P^\nu)1^*$, $C^*_3 = W^{*,2} = W^*_\mu \bigstar W^{*\mu}$, $W^* = W1^*$, $W^*_\mu = \epsilon^*{}_{\mu\alpha\beta\rho} \bigstar J^{\alpha\beta} \times\hat{} P^\rho$. The use of the isoexponentiation of the preceding section yields the iso-Lorentz isotransformations plus the isotranslations $x'^{*\mu} = x^{*\mu} + T^{*\mu}$ where $T^{*\mu} = k_\rho[\eta^{*\mu\rho} + [\eta^{*\mu\alpha}, P_\alpha]^*/1! + ...]$ (see p. 267 of [7-11] for details).

We note that isomomenta isocommute on isospaces over isofields. Still, they do not commute when projected on conventional spaces over a conventional field, thus confirming the non-linearity of the P^* (3.1) symmetry, as desired.

7.4-5. Isospinorial isosymmetry.

Besides the isotopies of rotations [7-7] and spin [7-8] [7-9], the third basic isosymmetry for relativistic hadronic mechanics is the isospinorial covering of the Lorentz-Poincarè-Santilli isosymmetry $ISP^* = SL^*$ (2.C) $\bigstar$ T^* which is necessary for the study of hadrons with spin 1/2 and was first introduced in the 1995 paper [7-12].

The ISP^* symmetry is constructed by: 1) Linearizing the second order iso-Casimir invariant of P^* (3.1); 2) Constructing the iso-Dirac equation in terms of iso-gamma and iso-Pauli matrices; 3) Formulating the iso-gamma an iso-Pauli matrices in terms of Bohm hidden variables; 4) Representing the generators in terms of the latter isomatrices; 5) Constructing the isoalgebra and isogroups of the isospinorial isosymmetry. Regrettably, we cannot review this derivation to prevent an excessive length of this chapter and refer interested readers to its original derivation [7-12]

We limit ourselves to recall that, according to relativistic quantum mechanics, a particle is technically represented by a unitary irreducible representation of the *IPS* = *SL*(2.*C*)× *T*. Consequently, a particle according to relativistic hadronic mechanics is an isounitary isoirreducible isorepresentation of *ISP** = *SL** (2.*C*) ★ *T**. The important point here is that the transition from ISP to ISP* causes an alteration of the intrinsic characteristics of particles and light, which are technically due to the isorenormalizations under non-potential/NSA interactions. These isorenormalizations imply the emergence of a new notion of particles called isoparticle, which was first introduced in the 1978 Harvard University memoir [3-5] under the name of mutation for the intent of studying, in due time, the expectation that the intrinsic characteristics of the proton, including its spin, may change in the transition from its conditions as the nucleus of the Hydrogen atom to its conditions in the core of stars and black holes.

7.4-6. Isoaxioms

In this section, we outline and upgrade the isoaxioms of special isorelativity [7-2]-[7-5] [3-9] [4-5] [4-19]-[3-21] in their projection in the space-time of the experimentalist assumed to be the conventional Minkowski space $M(x, \eta, 1)$ over the reals R by noting that, when said isoaxioms are formulated on the iso-Minkowskian space-time $M^*(x^*, \eta^*, 1^*)$ over the isoreals R^* they formally coincide with the conventional axioms of special relativity.

Despite the latter equivalence, the projection of the isoaxioms into our space-time implies physically different laws. Recall that the Minkowskian geometry represents a homogeneous and isotropic space in three dimensions. In contrast, the iso-Minkowskian isogeometry [7-13] represents a generally inhomogeneous space due to the local variation of the density and a generally anisotropic space due to the change of geometry with the change of direction.

Consequently, special relativity axioms do not need to identify the selected direction of motion. In contrast, such an identification is necessary for special isorelativity due to the indicated variation of physical characteristics with the variation of the space direction.

Under the above clarifications, special isorelativity can be defined as an axiom-preserving completion of special relativity on iso-Minkowski isospaces over isoreal isofields. Its universal isospinorial isosymmetry *ISP** characterizes uniquely and unambiguously the following isoaxioms first formulated in the 1983 paper [7-5] (systematically studies in the 1991 monographs [7-2] [7-3] and then upgraded in various publications) here expressed for the selected *k*-direction, e.g., that of the third space component,

ISOAXIOM I: The speed of light within (transparent) physical media is given by the locally varying speed: $C = \frac{c}{n_4}$.

ISOAXIOM II: The maximal causal speed within physical media is given by $V_{max,\,K} = c\frac{n_k}{n_4}$.

ISOAXIOM III: The addition of speeds within physical media follows the isotopic law:

$$V_{tot} = \left(\frac{v1.k}{n_k} + \frac{v_{2.k}}{nn_k}\right)/\left(1 + \frac{v_1 v_2}{c^2}\frac{n_4^2}{n_k^2}\right).$$

ISOAXIOM IV: The isodilation of time, the isocontraction of lengths, the isovariation of mass and the mass-energy isoequivalence (isorenormalization) within physical media follow the isotopic laws: $t'^*_k = \gamma^*_k\, t$, $\ell'^*_k = \gamma_k^{*,-1}\,\ell$, $m'^*_k = \gamma^*_k\, m$, $E^*_k = m\, V^2_{max,k}$, $m^*_k = m\, n^2_k / n^2_4$, where

$$\beta^*_k = \frac{v_3/n_k}{c/n_4}, \quad \gamma^*_k = 1/\sqrt{1-\beta^{*2}_k}.$$

ISOAXIOM V: The frequency shift within physical media follows the isotopic law (for null aberration) $\omega_{receiver}/\omega_{source} = 1/\gamma^*\,[1 - \beta^*\, isocos(\alpha^*)]$.

Note that the maximal causal speed of special isorelativity is no longer given by the speed of light c. It is given instead by Isoaxiom II because physical media are generally opaque to light, thus requiring a broader geometric notion derivable from the expression in $(k-4)$-space $dx^2_k/n^2_k - dt^2c^2/n^2_4 = 0$, which implies the invariance of $V_{max,\,K} = c\, n_k/n_4$.

In the next chapters, we shall review the validity of special isorelativity and related methods for all relativistic conditions of extended particles/wave packets in conditions of mutual overlapping or deep entanglement studied to date. In this chapter's next subsections, we shall review the apparent experimental verifications of the notion of isoparticles of Sect, 7.3.5, including that of isophotons, via verification of the fundamental predictions of special isorelativity.

It should be noted that certain applications, such as those in nuclear physics, special isorelativity, and hadronic mechanics, allowed the first known numerically exact and time-invariant representation of nuclear data [7-19] that could not be achieved in one century of studies via special relativity and quantum mechanics. The author believes the above exact and invariant representations are the most significant verifications of space-time isosymmetries and isorelativites, given their otherwise impossible advances to solve our alarming environmental problems.

7.4-7. Verification with local speeds of light

Special isorelativity has provided the first known invariant description of Lorentz's local speed of light within transparent media (Sect. 7.3.3), such as that in water for which we have the light speed $C = 200\ K\ m/s = c/n - 4 = 2c/3$, $n_4 = 3/2 = 1.5$. Note that, under the indicated reduced speed, the notion of isophoton solely applies for sufficiently big frequencies, while for all remaining speeds, the notion of isolight applies.

It is instructive for the reader to verify that the above notions of isophoton and isolight resolve Insufficiencies I to VII of Sect. 7.3.2. In fact, according to Isoaxiom III, the sum of two speeds of isolight $C = c/n4$ is $C = c/n4$. Moreover, water can be assumed homogeneous and isotropic with consequential simplified isometric $\delta^* = Diag.(1, 1, 1, 1/n^2_4)$. In this case, the maximal causal speed for massive particles in water is the speed of light c in a vacuum, therefore providing the first and only known geometric representation of the Kerenkov light according to which electrons travel in water with a fully causal speed bigger than the local light speed $C = 0.66\ c$.

It should be indicated that the above results are based on the experimental evidence that light is a "transverse electromagnetic wave" with quantized absorption rather than a quantized propagation for all frequencies as assumed by mainstream physics during the past century.

Intriguingly, the iso-Minkowskian geometry [7-13] provides a geometric representation of the difference between our view of the actual size duct of an object outside water and that an observer perceives the same object in water dint. For this purpose, note that we have the isounit I* = n24 = 9/4 > 1 in water. Then the use of the basic law for isotopy [4-5] $d^2act \times I^* = d^2_{int} \times I$, yields the value $d_{int} = n_4 d_{act} = 3/2\ d_{act}$ which is essentially verified by visual inspection, as one can see by looking at a rule partially submerged in water.

Intriguing similar properties occur for other external and internal characteristics. For instance, we have a similar distinction between our time t called external time t and related unit $I_t = 1$ (which is the time of the external observer), and the internal time t^* with related isounit $I^*_{t,}$ which is the time for an internal observer with interconnecting isotopic law $t \times It = t^* \star I^*_t$.

The notion of internal isotime appears to be significant for biological structures [4- 10] because we generally assume that the time sensed by a biological structure such as a seashell 1* is identical to our time t, while in reality, the external and internal times may be different with intriguing implications for the way biological structures may perceive our world.

7.4-8. Verification within local frequencies of light

The second important prediction of special isorelativity is that the intrinsic characteristics of particles and electromagnetic waves can be mutated in the transition from propagation in a vacuum to propagation within physical media, with the ensuing transition from particles and photons to isoparticles and isophotons, due to the new isorenormalizations caused by non-potential/NSA interactions.

The author has verified this second prediction via systematic experiments on light mutation into isolight. Recall that the Doppler law in a vacuum can be written in first-order approximation via the frequency shift $z = \omega'/\omega - 1 \approx \pm v/c$, where ω' (ω) denotes the frequency measured by the observer (at the source) and the minus (plus) sign occurs for the source moving away (toward) the observer.

The corresponding expansion for the Doppler-Santilli isolaw of Isoaxiom V yields the frequency shift $z^* = \omega'^*/\omega^* -1 \approx \pm v_k\ /c\ n_4\ /n_k$, where the factor n_4/n_k can be expanded in terms of $(c/v_k)d$, d (representing the travel of light in the medium) $n_4/n_k \approx 1 \pm \Xi(\rho,\ \tau,)\ c/v_k\ d$, where Ξ is a function of the local density, temperature and possibly other local characteristics of the medium considered.

The iso-Doppler shift law of Isoaxiom V then assumes the form in first-order approximation, called Santilli's isoshift (because solely identifiable via isomathematics), first derived in the 1991 Refs. [7-2] [7-3] $z^* = \omega'/\omega - 1 \approx \pm v/c \pm \Xi(\rho, \tau, ...)d$, which represents two independent frequency shifts:

1) The conventional Doppler's shift $\pm v/c$ due to relative motion between observer and source;
2) The isoredshift $z^* = -\Xi\ d$ (isoblueshift $z^* = +\Xi\ d$) due to loss (absorption) of energy $E = hv$ to (from) a gas at temperatures lower (bigger) than $0\ C$.

Note that the isoshift law will be used in Chapter 10 for our studies in cosmology.

Santilli conducted systematic tests in the former laboratories of Magnegas Corporation located at 150 Rainville Rd, Tarpon Springs, Florida, via the isoshift Testing Station, which consists of a front (rear) air-conditioned cabin containing a blue laser light (wavelength analyzers), the two cabins being interconnected by a 60 *ft* long tube containing air at 1, 000 *psi*. The air temperature was varied from −30 C to +200 C via commercially available cooling or heating means.

Systematic tests reported in the 2010 paper [7-20] established that a blue laser light loses (gains) energy, thus decreasing (increasing) its frequency when passing through air contained in the indicated tube at 1 000 ψ and −10 C (+150 C).

Since none of the measured frequency shifts occurred with any relative motion between the source and the observer, tests [7-20] established the existence of the isoshift as predicted in the 1991 monograph [7-3], namely, the loss of energy by light to the medium, with a consequential decrease of its frequency, by therefore illustrating that in the transition from light to isolight we may have not only a mutation of its speed but also of its structural characteristics.

Systematic additional measurements were conducted in the USA and Europe via the best available wavelength analysers by following Sunlight with a telescope from the Zenith to the horizon. These tests established that the redness of the Sun at the horizon is due to such a loss of energy to our atmosphere to cause the isoredshift of the blue light into the red light [7-21] - [7-23]. It should be noted that the widely assumed scattering of photons among the molecules of our atmosphere is insufficient to represent the large redshift from the blue to the red.

7.4-9. Verification via arbitrary speeds within physical media

The first known hypothesis of arbitrary causal speeds of massive physical particles under strong interactions was formulated by Santilli in the 1982 Harvard University paper [3-24] under DOE support. The hypothesis is based on the fact that the contact, non-potential/NSA component of the strong interactions between extended hadrons carries no energy, a feature first proved in Table 5.1, p. 827 on of the 1978 memoir [3-5] and recently confirmed in paper [7-25]. As it is well known, achieving the speed *c* for massive particles under sole action-at-a-distance potential interactions requires infinite energy. In contrast, the same particles under contact non-potential interactions would accelerate beyond *c* without any violation of the principle of conservation of energy.

Subsequently, Santilli then achieved in the 1983 paper [7-5] the universal invariance $SO(3.1)^*$ of arbitrary speeds of light by proving, in particular, its isomorphism with the Lorentz symmetry $SO(3.1)$ and the admission of arbitrary speed [7-25] [7-26]. The same result can also be achieved via the general relativity covariance $x^\mu \rightarrow x^{*\mu} = x^\mu / n_\mu$, $\mu = 1, 2, 3, 4$ that, when formulated in isospaces over isofields, preserves all axioms and physical laws of special relativity, yet the local speed of light in the space-time of the experimenter is $C = c/n_4$.

Subsequent studies proved the compatibility of arbitrary speeds with the abstract axioms of special relativity [7-25] [7-26]. A comprehensive study is available in monographs [4-4] [4-5].

To the author's best knowledge, by ignoring scientifically unfounded views, the above findings have been experimentally verified as follows:

7.4-10. Verifications in hadronic structures

By noting that hadrons have hyperdense interior media with an ensuing lack of exact validity of quantum mechanics (Chapter 3), H. P. Nielsen and I. Picek [7-27] have shown the universality of the exact fit of experimental data for the interior of hadrons by hadronic mechanics beginning with pions, with ensuing maximal causal speeds bigger than c. This feature is crucial for quantitatively representing the synthesis of the neutron from the

Hydrogen and related new recycling of radioactive nuclear waste [7-19], outlined in the next chapter.

V. de Sabbata and M. Gasperni [7-28] showed via the use of conventional gauge theories that the maximal causal speed within hyperdense hadronic media can be as big as 75 c (see Appendix B, p. 306 on of monograph [4-19] for a detailed review). Additional studies were done by E. Recami and R. M. Santilli [7-29] to clarify that the superluminal speeds here considered do not refer to tachyons. More recently, Santilli [7-19] achieved the first identification of the neutron's density, resulting in smaller than one, $n_4 < 1$. The ensuing maximal causal speed in the interior of the neutron $C = c/n_4 > c$ resulted in being necessary for a numerically exact and time-invariant representation of all characteristics of the neutron in its synthesis from the Hydrogen in the core of stars (see Chapter 8 for details).

7.4-11. Verifications in astrophysics

Many astrophysical measurements (see, e.g., Refs. [7-30]-[7-32]) have indicated the expulsion of masses from astrophysical bodies at speeds bigger than the speed of light c. These measurements have been rejected by mainstream academia, again, on theoretical grounds of violation of special relativity, while said measurements are fully compatible with special relativity under the admission of contact non-potential interactions causing the expulsion of large astrophysical masses and their representation via isomathematics.

7.4-12. Verification in neutrino Experiments

The neutrino speed measurements while traveling underground were conducted at CERN from 2008 to 2011 by a considerable number of experimentalists as part of the OPERA project that included the detection of CERN-emitted neutrinos at the Gran Sasso Laboratory in Italy. The resulting superluminal neutrino speeds were reported in the CERN preprint dated November 17, 2011 [7-33], submitted to the Journal of High Energy Physics on the same date.

Despite the credibility of the laboratory, despite the number of committed experimentalists, despite the systematic conduction of the measurements over three years, despite considerable supporting research on superluminal speeds within physical media [7-34]-[7-36] and despite considerable interest by society [7-37], the Journal of High Energy Physics rejected the paper via theoretical elaborations of form factors and renormalizations that, in the view of various physicists, appeared to be aimed at adapting experimental data to special relativity.

In the author's view, special relativity cannot be credibly claimed to be exactly valid through the 557 miles of underground rocks and soil between CERN and the Gran Sasso Laboratory. Hence, publicly supported research in the field should aim at generalizing special relativity for the considered interior conditions rather than adapting physical reality to a century-old preferred theory for point particles in a vacuum.

Moreover, the very existence of neutrinos is in serious doubt because relativistic quantum mechanics requires that, to verify the principle of conservation of energy, in the nuclear fusion of two nuclei into a third, the total rest energy of the two original nuclei has to be bigger than that of the third nucleus, resulting in an effect at the foundation of nuclear energy called mass defect.

By comparison, synthesizing the neutron from the Hydrogen in the core of stars $p^+ + e^- \rightarrow n + \nu$, we have the original rest energy $E_p + E_e = 938.272 + 0.511 MeV = 938.783\ MeV$. In contrast, the resulting neutron has the rest energy of $E_n = 939.565\ MeV$, as a result of which the synthesis of the neutron requires $\Delta_E = E_n - (E_p + E_e) = 0.782\ MeV > 0$, resulting in a mass excess. Therefore, by noting that there is insufficient energy to create the neutron, under the assumption of special relativity and quantum mechanics, the century-old theoretical assumption of the emission of a hypothetical neutrino ν in the neutron synthesis from the Hydrogen violates the principle of conservation of the energy (see Chapter 8 for details).

7.4.-13 Lie-admissible relativities

As it is well known, symmetries and conservation laws constitute an important part of 20th century physics because Lie symmetries best characterize conservation laws. As a simple example, the invariance of a Hamiltonian $H\ (r,\ p)$ under the one dimensional Lie symmetry $H(t) = exp\{Hti\}H(0)exp\{-it\mathrm{H}\}$ implies, for infinitesimal values of t, Heisenberg's time evolution $i\ dH/dt = [H, H] = HH - HH \equiv 0$, and similar results hold for other continuous or discrete symmetries.

By recalling that Lie-admissible groups can also leave invariant physical quantities, the primary aim of Lie-admissible symmetries and relativity is characterizing the time rate of variation of physical quantities. Suppose, for instance, that the Hamiltonian explicitly depends on time, $H = H(t, r, p)$. Then it is invariance under the one dimensional Lie-admissible group $H(t) = exp\{HSti\}H(0)exp\{-itRH\}$, where R and S are the backward and forward (dimensionless) Santillians, respectively, implies, for infinitesimal values of t, the Lie-admissible time evolution $i\ dH/dt = (H, H) = HRH - HSH = H(R - S)H$ which, for $R = 1$ and $S = 1 - F/H$, yields $i\ dH/dt = (H, H) = H F = 0$, where F represents Lagrange's and Hamilton's external forces responsible for time irreversibility. By keeping in mind the universality of the Lie-admissible formulations for all possible (non-singular) non-conservative classical systems shown in Sect. 7.2, similar results hold for the remaining components of the classical Lie-admissible relativity introduced in the 1978 Harvard memoir [3-4].

To the author's best knowledge, no additional study on Lie-admissible relativities is on record at this writing (December 10, 2024), except the proof of the universality of Lie-admissible formulations for all possible (non-singular) non-conservative operator systems such as nuclear fusions or combustion [7-39].

7.5 Santilli's general isorelativity

There is no doubt that Einstein's general relativity has the historical privilege of having initiated the geometric study of the exterior gravitational problem, namely, the representation via the Riemannian geometry of the gravitational field generated by a massive body [7-40].

Despite that, in Santilli's view, general relativity is one of the most controversial physical theories because of the lack in qualified scientific journals over one century of due process on any of the following open problems (see ref. [7-17] for details):

7.5.1. Lack of characterization of space as a prerequisite for the physical meaning of the curvature of space, since the curvature of the generally assumed empty void cannot possibly guide masses in curved trajectories.

7.5.2. Lack of physical realization of the Riemannian curvature in three dimensions. Since, as it is well known, the curvature of space can solely be realized in two dimensions, thus creating doubts as to whether such a physically unknown curvature can move planets in their orbits around Earth.

7.5.3. Lack of invariance over time of the numerical predictions of general relativity. This insufficiency is due to the incompatibility of Lie's theory with the Riemannian geometry with ensuing lack of symmetries in general relativity and the replacement of the invariance of special relativity with covariance, as a result of which the experimental verifications of general relativity solely hold at one fixed point in time.

7.5.4. Incompatibility of general relativity with special relativity, as established, e.g., by the impossibility of obtaining the ten conservation laws of the Poincare' symmetry at the limit of null Riemannian curvature.

7.5.5. Incompatibility of the Riemannian geometry with the physical notion of waves, by therefore casting shadow on the claimed detection of gravitational waves.

7.5.6. Incompatibility of general relativity with quantum mechanics, e.g., because, on analytic grounds, general relativity is a non-canonical theory. Consequently, the only possible operator image of general relativity is a non-unitary theory with ensuing problematic aspects studied in Chapter 5, including the possible violation of causality.

7.5.7. Incompatibility of general relativity with gauge theories, since general relativity is a classical non-canonical theory while gauge theories are operator unitary theories with ensuing impossibility of reaching a grand unified theory known since Einstein's times.

7.5.8. Incompatibility of the Riemannian geometry with the interior gravitational problem due to the basically insufficient functional dependence of the Riemannian metric for interior gravitational conditions.

7.5.9. Incompatibility of general relativity with antimatter, evidently due to the lack of a consistent inclusion of the needed anti-Hermitean conjugation.

In Santilli's view, all problematic aspects 7.5.1 to 7.5.10 of general relativity are due to the representation of gravitation via the physically unknown curvature of a physically unknown space, with the consequential loss of Lie's theory and related invariance. In the hope of initiating a due scientific process for the resolution of the insufficiencies, Santilli [7-15] proposed at the 1992 Seventh Marcel Grossmann Meeting on Gravitation held at SLAC, California, the formulation of Einstein's field equations and related general relativity on the iso-Minkowski isospace M^* with isometric η * over the field of isoreal isonumbers R^* under the name of general isorelativity, or isogravitation for short, via the following steps:

1) Decomposition of the Riemannian metric g(x) into the product of the gravitational Santillian $S_g r$ and the Minkowski matric *η*, *g(x)* = $S_g r$ *η*.
2) Assumption of the isounit 1* = 1/ $S_g r$ as the multiplicative unit of the theory with consequential isotopic formulation of the mathematical machinery of the Riemannian geometry [7-13].
3) Use of the unrestricted functional dependence of the iso-Minkowskian metrics *η** = *η**(x, v, d, ...)* and, therefore, of the gravitational Santillian.

The deformation of the Minkowski metric into the Riemannian metriic $\eta \rightarrow$ g(x) = S *η* while jointly deforming the unit of the Riemannian geometry into the opposite of the gravitational Santillian, $1 \rightarrow 1^* = 1/S_{gr}$, implies that the iso-Minkowskian geometry is isoflat, namely, it has zero curvature when formulated on isospace *M** over the isofield *R** while recovering Einstein's curvature in full when projected on the Riemannian space over the field of real numbers.

Thanks to the compatibility of the iso-Minkowskian geometry with the Lie-isotopic theory, the above assumptions restore the notion of invariance in lieu of covariance, with ensuing invariance over time of the numerical predictions thanks to the invariance of Riemannian line elements under the iso-Poincare' isosymmetry [7-11].

The compatibility of isogravitation with special relativity is uniquely and unambiguously established by the reduction of the iso-Poincare' symmetry to the conventional Poincare' symmetry under the limit of its isounit $1^* \rightarrow 1$.

The restoration of the physical notion of waves for a gravitational field is established by the applicability of Maxwell's equation to the iso-Minkowskian geometry.

In the 1997 Eight Marcel Grossmann Meeting on Gravitation held in Jerusalem, Israel, Santilli [7-16] showed that the causal, unique and unambiguous operator version of isogravitation is given by relativistic hadronic mechanics when its isounit is the inverse of the gravitational Santillian 1* = 1/.$_{Sgr}$. The topological equivalence of the iso-Minkowskian and Minkowskian geometries then implies the compatibility of isogravitation with gauge theories.

An isotopic model of grand unification of gravitation and electroweak interactions was finally achieved in monograph [4-11] following the first known representation of the gravitational field of antimatter.

Following the forgotten second paper by Schwartzschild [7-41] because incompatible with the exterior character of general relativity, and thanks to the arbitrary functional dependence of the iso-Minkowski metric $\eta(x, v, d, ...)$, isogravitation has allowed the first known quantitative study of the interior gravitational problem here intended as the study of the origin (rather than the description) of gravitation which is studied in the next chapter following the reduction of neutron to a hadronic bound state of a protons and an electron with ensuing verification of the Poincarè hypothesis according to which the exterior gravitation of a massive body appears to be entirely reducible to the electromagnetic interactions of its electron and proton constituents.

In closing this chapter, we hope to have shown that special and general isorelativities are mere broader realizations of the abstract axioms of Einstein's special relativity with the consequential dramatic broadening of their application from point-like particles in vacuum to extended particles within physical media as well as to interior and exterior gravitation.

In closing, we should recall that conventional relativities are solely valid when considered in an inertial reference frame while, by contract, Li-isotopic and Lie-admissible relativities are also valid in non-inertial reference frames in view of the arbitrary functional dependence of the Santillian S, thus including a dependence on accelerations.

References of Chapter 7

[1] Muktibodh, A. S. and Santilli, R. M.: Studies of the Regular and Irregular Isorepresentations of the Lie-Santilli Isotheory, Journal of Generalized Lie Theories 11, 1-7 (2007), http://www.santilli-foundation.org/docs/isorep-Lie-Santilli-2017.pdf

[2] Santilli, R. M.: Isotopic Generalizations of Galilei and Einstein Relativities, International Academic Press (1991), Vol. I Mathematical Foundations, http://www.santilli-foundation.org/docs/Santilli-01.pdf

[3] Santilli, R. M.: *Isotopic Generalizations of Galilei and Einstein Relativities, International Academic Press* (1991), Vol. II: Classical Formulations, www.santilli-foundation.org/docs/Santilli-61.pdf

[4] Jannussis, A.: Nomination of R. M. Santilli for the 1992 Nobel Prize in Physics, Link to be added

[5] Santilli, R. M.: Lie-isotopic Lifting of Special Relativity for Extended Deformable Particles, *Lettere Nuovo Cimento* 37, 545-555 (1983), www.santilli-foundation.org/docs/Santilli-50.pdf

[6] Santilli, R. M.: Lie-isotopic Lifting of Unitary Symmetries and of Wigner's Theorem for Extended and Deformable Particles, *Lettere Nuovo Cimento* 38, 509-521 (1983), www.santilli-foundation.org/docs/Santilli-51.pdf

[7] Santilli, R. M.: Isotopies of Lie Symmetries, I: Basic theory, *Hadronic* J. 8, 8-35 (1985), II: Isotopies of the rotational symmetry, *Hadronic* J. 8, 36-52 (1985), www.santilli-foundation.org/docs/santilli-65.pdf

[8] Santilli, R. M.: Rotational isotopic symmetries, ICTP communication No. IC-91-261 (1991), www.santilli-foundation.org/docs/Santilli-148.pdf

[9] Santilli, R. M.: Isotopic Lifting of the *SU*(2) Symmetry with Applications to Nuclear Physics, *JINR Rapid Comm*. 6. 24-38 (1993), http://www.santilli-foundation.org/docs/Santilli-19.pdf

[10] Santilli, R. M.: Lie-isotopic generalization of the Poincare' symmetry, classical formulation, ICTP communication No. IC/91/45 (1991), www.santilli-foundation.org/docs/Santilli-140.pdf

[11] Santilli, R. M.: Non-linear, Non-local and Noncanonical Isotopies of the Poincare' Symmetry, *Moscow Phys. Soc*. 3, 255-239 (1993), www.santilli-foundation.org/docs/Santilli-40.pdf

[12] Santilli, R. M.: Isotopies of the spinorial covering of the Poincare symmetry,' (*Chinese J. System Eng. and Electr*. 6, 177-199 (1995), www.santilli-foundation.org/docs/Santilli-18.pdf

[13] Santilli, R. M.: Isominkowskian Geometry for the Gravitational Treatment of Matter and its Isodual for Antimatter, *Intern. J. Modern Phys*. D7, 351-407 (1998), www.santilli-foundation.org/docs/Santilli-35.pdf

[14] Aringazin, A. K. and Aringazin, M.: Universality of Santilli's iso-Minkowskian geometry in *Frontiers of Fundamental Physics*, M. Barone and F. Selleri, Editors, Plenum (1995), http://www.santilli-foundation.org/docs/Santilli-29.pdf

[15] Santilli, R. M.: Isotopic quantization of gravity and its universal isopoincarè symmetry, in the Proceedings of *The Seventh Marcel Grossmann Meeting on Gravitation*, SLAC, 1992, R. T. Jantzen, G. M. Keiser and R. Ruffini, Editors, World Scientific Publishers, 500-505 (1994), www.santilli-foundation.org/docs/Santilli-120.pdf

[16] Santilli, R. M.: Unification of gravitation and electroweak interactions, in the Proceedings of the *Eight Marcel Grossmann Meeting on Gravitation*, Israel, 1997, T. Piran, and R. Ruffini, Editors, World

Scientific, 473-475 (1999), www.santilli-foundation.org/docs/Santilli-137.pdf

[17] Pfapt, P.: Einstein's General Relativity or Santilli's Iso-Relativity?, website, www.eprdebates.org/docs/epr-review- ii.pdf

[18] Santilli, R. M.: Compatibility of Arbitrary Speeds with Special Relativity Axioms for Interior Dynamical Problems, *American Journal of Modern Physics*, 5, 143 (2016), www.santilli-foundation.org/docs/ArbitrarySpeeds.pdf

[19] Santilli, R. M., Guest Editor, *Lie-isotopic representation of nuclear data*, Special Issue of *Ratio Mathematica* 52 (2024), http://www.santilli-foundation.org/docs/RM-Special-Issue-.Vol.-52-2024-compressed.pdf

[20] Santilli, R. M.: Experimental Verifications of IsoRedShift with Possible Absence of Universe Expansion, Big Bang, Dark Matter, and Dark Energy, *The Open Astronomy Journal* 3, 124 (2010), http://www.santilli-foundation.org/docs/Isoredshift-Letter.pdf

[21] Santilli, R. M., West, W. and Amato, G.: Experimental Confirmation of the IsoRedShift at Sunset and Sunrise with Consequential Absence of Universe Expansion and Related Conjectures, *Journal of Computational Methods in Sciences and Engineering*, 12, 165-182 (2012), http://www.santilli-foundation.org/docs/Confirmation-sun-IRS.pdf

[22] West, W. and Amato, G.: Experimental Confirmation of Santilli's IsoRedShift and IsoBlueShift, lecture delivered at the San Marino Workshop on Astrophysics and Cosmology for Matter and Antimatter, Republic of San Marino (2011), www.world-lecture-series.org/san-marino-2011-exp-conf-Santilli-isoredshift-iso blueshift

[23] Ahmar, H., Amato, G., Kadeisvili, J. V., Manuel, J., West, W. and Zogorodnia, O.: Additional experimental confirmations of Santilli's IsoRedShift and the consequential lack of expansion of the universe, *Journal of Computational Methods in Sciences and Engineering* 13, 321-342 (2013), http://www.santilli-foundation.org/docs/IRS-confirmations-212.pdf

[24] Santilli, R. M.: Studies on A. Einstein, B. Podolsky and N. Rosen prediction that quantum mechanics is not a complete theory, III: Illustrative examples and applications, *Ratio Mathematica* 38, 139-222 (2020), http://eprdebates.org/docs/epr-review-iii.pdf

[25] Santilli, R. M.: Universality of special isorelativity for the invariant description of arbitrary speeds of light, Nova collection of papers dedicated to the Lorentz symmetry, V. Dvoeglazov, Editor, Nova Publisher (1997), arXiv: physics/9812052 (1997)

https://arxiv.org/pdf/physics/9812052.pdf

[26] Santilli, R. M.: Compatibility of Arbitrary Speeds with Special Relativity Axioms for Interior Dynamical Problems, *American Journal of Modern Physics* 5, 143-154 (2016), http://www.santilli-foundation.org/docs/ArbitrarySpeeds.pdf

[27] Nielsen, H. P. and Picek, I.: Nuclear Physics B-211, 289 (1983), Link to be provided

[28] de Sabbata, V. and Gasperini, M.: *Lettere Nuovo Cimento* 34, 337-349 (1982), Link to be provided

[29] Recami, E. and Santilli, R. M., Experimental evidence of superluminal speeds, *Hadronic Journal* 23, 279-294 (2000), https://www.santilli-foundation.org/docs/santilli-502.pdf

[30] Mirabel, F. and Rodriguez F., superluminal source in the Galaxy, *aNute* 371, 364 (1994), https://www.nature.com/articles/371046a0

[31] Tingay, J. t al, Relativistic motion in a nearby bright X-ray source, *Nature* 374, 141 (1996), https://www.nature.com/articles/374141a0

[32] Gianni, S., Experimental evidence of superluminal velocities in astrophysics and proposed experiments AIP *Conf. Proc*. 458, 881–888 (1999) https://pubs.aip.org/aip/acp/article-abstract/458/1/881/580805/Experimental-evidenceof-superluminal-velocities

[33] A. Ereditato et al.: Measurement of the neutrino velocity with the OPERA detector in the CNGS beam, CERN preprint, http://www.santilli-foundation.org/docs/OPERA-experiment.pdf

[34] Brumfiel, G. Particles break the light-speed limit. *Nature*, September 22 (2011)., https://doi.org/10.1038/news.2011.554

[35] A. Enders and G. Nimtz, Photonic-tunneling experiments *Phys. Rev.* B 47, 9605- 9609 (1993), journals.aps.org/prb/abstract/10.1103/PhysRevB.47.9605

[36] Workshop on Astrophysics and Cosmology for Matter and Antimatter, Republic of San Marino, September 5 to 9, 2011 (2011), website http://www.workshops-hadronic-mechanics.org/workshop-results-matter.php

[37] BBC news release: Neutrino experiment repeat at Cern finds same results (2012), http://www.bbc.co.uk/news/science-environment

[38] Santilli, R. M.: The etherino and/or the neutrino Hypothesis? *Found. Phys*. 37, 670-695 (2007), http://www.santilli-foundation.org/docs/EtherinoFoundPhys.pdf nt-15791236

[39] Santilli, R. M.: Lie-admissible invariant representation of irreversibility for matter and antimatter at the classical and operator

levels. *Nuovo Cimento* B 121, 443-485 (2006) //www.santilli-foundation.org/docs//Lie-admiss-NCB-I.pdf

[40] Schwartzschild, K.: Uber das Gravitationsfeld eines Massenpunktesnack der Einsteinshen Theorie, Sitzber. Deut. Akad. Wiss. Berlin, Kl. Math.-Phys. Tech., 189-196 (1916).

[41] Schwartzschild, K.: Uber das Gravitationsfeld einer Kugel aus inkompressibler Flussigkeit nach Einsteinshen Theorie, Sitzber. Deut. Akad. Wiss. Berlin, Kl. Math. - Phys. Tezh, 424-434 (1915)

CHAPTER 8

APPLICATION OF HADRONIC MECHANICS TO NUCLEAR PHYSICS

8.1. Insufficiencies of quantum mechanics in nuclear physics

In one century of research under large public funds, nuclear physics has been unable to solve the following basic problems:

8.1A. A quantitative representation of the structure of the neutron following its synthesis from the permanently stable proton and electron in the core of stars despite their extremely big Coulomb attraction of 230 Newtons at the mutual distance of the neutron size.

8.1B. The exact representation of the experimental data of stable nuclei, such as nuclear magnetic moments and spin, for nuclei in the ground states of the experimental detection that, by definition, are nuclei with null orbital contributions.

8.1C. Nuclear stability despite the natural instability of the neutron and extremely repulsive Coulomb protonic forces.

8.1D. The recycling of radioactive nuclear waste, e.g., via its stimulated decay.

8.1E. Controlled and sustainable nuclear fusions at large, and those of light stable nuclei in particular, which are predicted to produce nuclear energies without harmful radiations or radioactive waste.

In our view, the inability in one century to achieve a consistent quantitative formulation of the above problems, let alone their solution, is clear evidence of the insufficiencies of quantum mechanics in nuclear physics, with particular reference to the following insufficiencies [1-3]:

8.1-1. Inapplicability of the Schrödinger equation

The validity of the Schrödinger equation for a quantum mechanical (qm) bound state of an electron with mass $m_e = 0.511$ MeV and a proton with mass $m_p = 938.272$ MeV, relative coordinate r, reduced mass $\mu = m_p m_e/(m_p + m_e)$ and Coulomb potential $V_c = -e^2/r$ (for h= 1, c = 1) $[(1/2\mu)\Sigma_{k=1,2,3} p_k p_k - e^2/r]|\psi(r)\rangle = E|\psi(r)\rangle$ is justly set in history because of the exact and invariant representation of all experimental data of the Hydrogen atom $H = (p^+, e^-)_{qm}$. Nevertheless, the same equation is *inapplicable* for the synthesis of the neutron from a proton and an electron in the core of stars because said equation prohibits any relative distance smaller than Bohr's orbit $r = 10^{-8}$ cm, against the experimentally measured value for the neutron radius $R_n = 0.87 \times 10^{-13}$ cm. In Santillli's view, this insufficiency is due to the *locality* of quantum mechanics here referred to as the point-like abstraction of particles, which is acceptable for atomic structures due to the large relative distances between the constituents but excessive for the neutron synthesis due to the total mutual penetration and entanglement of the wave packets, with ensuing contact non- potential interactions whose description requires a structural generalization of quantum mechanics.

8.1-2. Inapplicability of Heisenberg's uncertainties

The validity of Heisenberg's uncertainty principle under the conditions of its original conception and experimental verification, i.e., for point-like particles in a vacuum under sole potential interactions, is beyond scientific doubt. Nevertheless, the *inapplicability* of the same principle for extended wave packets in total mutual entanglement is equally beyond scientific doubt, e.g., because any representation of the latter conditions requires a *non-unitary* generalization of quantum mechanics under which Heisenberg's uncertainty principle cannot be formulated, let alone be tested. At any rate, the experimental values of the neutron synthesis violate Heisenberg's uncertainties (Chapter 6) because the standard deviations of the electron coordinate Δr are much bigger than the size of the neutron $R_n = 0.87\ 10^{-}13$ cm and the standard deviations of the momentum imply that the electron has a kinetic energy inside the star bigger than the neutron's rest energy, therefore prohibiting any quantitative treatment of the neutron synthesis.

8.1-3: Inapplicability of Planck's quantum theory

The validity of energy exchanges in atomic structures according to Planck's quantum theory $E = h\nu$ for point-like particles in vacuum under sole potential interactions is also beyond scientific doubt. Nevertheless, the *inapplicability* of Planck's quantum theory for energy exchanges under strong interactions is also beyond scientific doubt because, as reviewed in Chapter 2, non-unitary transformations needed to represent the contact non-potential interactions in deep mutual entanglements of particles must be applied to the *totality* of the quantum formalism, beginning with the non-unitary transformation of the quantum mechanical unit h= 1 into Santilli's integro-differential isounit for strong interactions, $h = 1 \rightarrow 1^* = U h U^\dagger$ without any general possibility of discrete values. Nuclear fusions provide the first illustration of this occurrence. As it is well known, quantum mechanics can only describe the fusion of two particles with energies E_1 and E_2 into a third particle with energy E_3 *smaller* than the sum of the original masses, resulting in the known *mass defect* $\Delta E = (E_1 + E_2) - E_3$ which is at the origin of nuclear energies. The important point here is that, as established by tabulated data, *mass defects in nuclear fusions cannot be generally quantized.* A second illustration is provided by tabulated *nuclear decays,* such as beta decays, whose energy releases cannot be quantized. The neutron synthesis provides a third illustration. By recalling the values of rest energies $E_e = 0.511$ MeV, $E_p = 938.272$ MeV and $E_n = 939.565$ MeV, the synthesis of the neutron from the Hydrogen implies the *mass excess* of 0.782 MeV $= E_n - (E_e + E_p)$, namely, the neutron synthesis in a star *requires,* rather than releases, the un-quantizable energy of 0.782 MeV, and the same holds for all syntheses of unstable particles from lighter physical constituents [7-18], whose origination is one of the most mysterious events in nature (see Ref. [8-29]).

In this chapter, we show that, by assuming the completion of Planck's quantum $h = 1$ into the integro-differential isounit $1^* = U h U^\dagger = 1/S > 0$ with ensuing completion of all quantum products into the isoproduct $AB \rightarrow A \star B = ASB$ and the representation of all non-linear, non-local and non-potential forces via Santillians of the type used for isosymmetries and isorelativities (Chapter 6), $S = \mathrm{Diag.}(1/n_1^2, 1/n_2^2, 1/n_3^2, 1/n_4^2) \times e^{-\Gamma} > 0$ $S \ll 1$, hadronic mechanics allows numerically exact and time-invariant, non-relativistic and relativistic representations of *all* characteristics of the neutron in its synthesis from the electron and the proton, as well as of nuclear stability despite the neutron instability and the strongly repulsive Coulomb protonic forces, with ensuing new vistas for recycling of radioactive nuclear waste and nuclear fusions of light natural elements without harmful radiations or waste.

Similarly, in view of the very small area covered by lightning with ensuing prevention to interpret thunder as a chemical or dynamical process, the sole known quantitative representation of thunder is that lightning synthesizes Nitrogen and other elements from atmospheric components without any production of harmful radiations or release of radioactive waste.

8.2. The synthesis of neutroids and neutrons from the Hydrogen

8.2.1 Historical notes

As it is well known, stars initiate their lives as an aggregate of Hydrogen that grows by accretion during travel in interstellar spaces. At the moment when the temperature in the core of the aggregate reaches a value of the order of 10M K, E. Rutherford [8-1] suggested in 1920 that the Hydrogen atom is "compressed" into a new neutral particle, which he called the *neutron,* $e^- + p^+ \rightarrow n$.

The existence of the neutron was experimentally established in 1932 by J. Chadwick [8- 2]. In 1933, W. Pauli [8-3] noted that Rutherford's synthesis of the neutron violates special relativity because of the lack of conservation of the angular momentum since the proton and the electron have spin 1/2 and are expected to have a singlet coupling (antiparallel spin) resulting in a new particle with spin zero while the neutron has spin 1/2.

Therefore, to regain the validity of special relativity, Pauli suggested in 1933 the hypothesis that the neutron is synthesized jointly with the *emission* of a massless and charge-less particle with spin 1/2 that was called in 1935 by E. Fermi [8-4] the *neutrino* (meaning "little neutron" in Italian), according to the synthesis $e + p \rightarrow n + \nu$ which hypothesis has remained dominant for the neutron synthesis to this day.

Subsequent experiments (see the recent review [8-5]) established that *the neutron is naturally unstable* with a mean life of $\tau = 877$ s and spontaneous decay $n \rightarrow e^- + p^+ + \bar{\nu}$, where $\bar{\nu}$ is Fermi's *antineutrino.*

Predictably, the synthesis of the neutron from the Hydrogen attracted attention soon following Chadwick's confirmation. According to the historical account by R. Norman and J. Dunning-Davies [8-6], Ernest J. Sternglass conducted in 1951 the first test for the laboratory synthesis of the neutron from the Hydrogen, followed by tests in 1952 by E. Trounson and others, although none of these initial tests were published in view of the incompatibility of the neutron synthesis with quantum mechanics and for other reasons.

8.2.2. Don Borghi neutroid

To the author's knowledge, the first published tests on the laboratory synthesis of the neutron from Hydrogen were conducted in the 1960's by the Italian priest-physicist Don Carlo Borghi of the University of Milan, Italy, and his associates [8-7]. In essence, the experimentalists constructed a cylindrical metal chamber (called a klystron) filled with Hydrogen gas (at a pressure of about 1 bar) that was kept mostly ionized via an electric arc with approximately 500 V and 10 mA. Additionally, the gas was traversed by microwaves with a frequency of 10^{-10} s.

The experimentalists then placed various materials suitable for activation when exposed to a neutron flux (such as gold or silver) on the exterior of the klystron. Following exposures over several weeks, the experimentalists reported clear and reproducible nuclear transmutations that can only be attributed to a neutral particle emitted from the klystron.

Due to insufficient knowledge on neutron emission at the time, the experimentalists conjectured that the detected nuclear transmutations were due to a new, neutral, strongly interacting particle with the mass of the neutron that they call the *neutroid* (Figure 8-1).

8.2.3. Santilli's isoneutron

As it is well known, the Hydrogen atom H is a quantum mechanical (qm) bound state of an electron and a proton $H = (e^-, p^+)_{qm}$. Santilli submitted in the 1978 Harvard University memoir [3-5] and then studied in Refs. [8-8]-[8-15] at the theoretical level and Refs. [8- 15]-[8-21] at the experimental level (see also Refs. [8-22]-[8-26] by independent scholars,) the structure model of the neutron n as a bound state according to hadronic mechanics (hm) of an electron and a proton in conditions of total mutual penetration and entanglement, resulting in the mutated particles nowadays called the isoelectron e^{*-} and the isoproton p^{*+} according to the structure model $n = (p^{*+}\uparrow, e^{*-\nu}\downarrow)_{hm}$ called *Santilli isoneutron*, where the prefix "iso" stands to indicate that the model is solely treatable via isomathematics.

Said model was submitted to represent, in first approximation, a permanently stable neutron so as to use the Lie-isotopic branch of hadronic mechanics at both the non-relativistic and relativistic levels, the extension to irreversible Lie-admissible treatments due to the natural instability of the neutron being done via the rules of Sect. 3.5.4.

The model was also submitted to represent the electron as being totally "compressed" [8-1] inside the proton by, therefore, having a *constrained angular momentum* L_e *equal to the spin of the proton* $J_p = 1/2$ to prevent extreme resistive forces. It then follows that, by recalling the singlet coupling of the structure, the total angular momentum J_e of the electron is null $J_e = S_e — L_e = 1/2 — 1/2 = 0$, and the spin of the neutron coincides with the spin of the proton $J_n = J_p — J_e + L_e = 1/2 = J_p$.

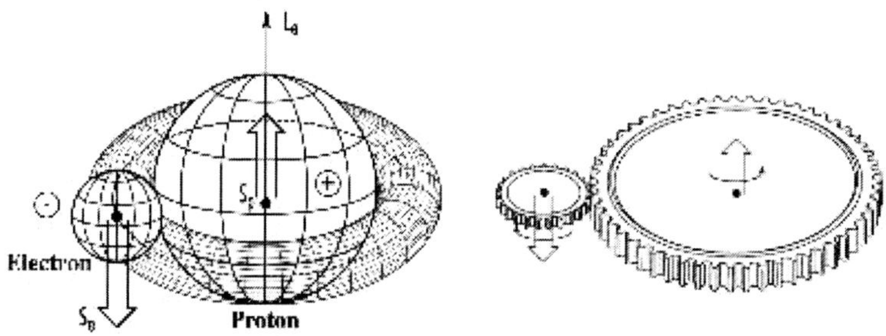

Figure 8.1. We depict the bound state predicted by hadronic mechanics of an electron and a proton in singlet coupling under the extremely big Coulomb attraction of 230 Newtons at 10^{-13} cm mutual distances represented by the Hamiltonian (left view) as well as the strongly attractive interactions due to contact non-potential interactions represented by the Santillian (right view), resulting in a new neutral particle with essentially the same mass, radius and mean life of the neutron but with spin zero, which is the best candidate for Don Borghi's neutroid (Sect. 8.2.2).

Since the electron mass is 5×10^{-4} smaller than the mass of the proton, the proton can be considered to be at rest, and un-mutated $n = (p^{*+}, e^{*,-})_{hm} \approx (p^{+}, e^{*,-})_{hm}$, and all isorenormalizations caused by non-Hamiltionian interactions can be restricted to those of the electron.

In the following, we show that the treatment via hadronic mechanics of the indicated model provides a numerically exact representation of all characteristics of the neutron, which is invariant over time due to the invariance of the theory under the iso-Galilean symmetry at the non-relativistic level and iso-Lorentz symmetry at the relativistic level (Chapter 7).

8.2.4 Representation of the neutron mass, charge radius and mean

Let $V_c = -e^2/r$ be the attractive Coulomb potential between the electron and the proton. The well-known reduced form of the Schrödinger equation for the Hydrogen atom is then given by $[(1/2\mu)d/dr(r^2d/dr) — e^2/r]\psi(r) = E\psi(r)$ where E is the electromagnetic binding energy.

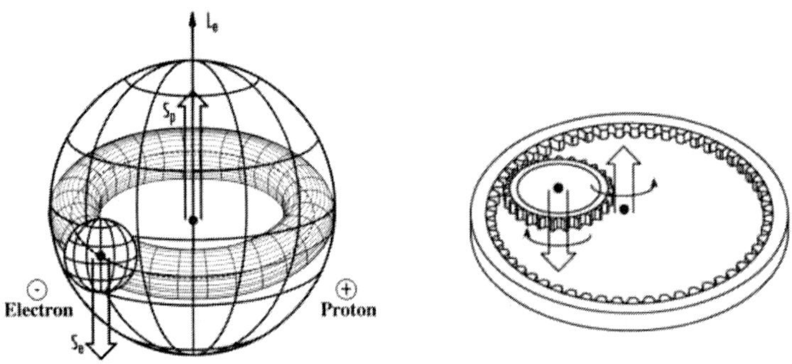

Figure 8-2. We illustrate the structure of the neutron according to Rutherford's [8-1] total "compression" of the electron within the hyperdense proton in the core of stars, under the potential Coulomb attraction illustrated on the left and the contact Santillian interaction illustrated on the right. As reviewed in this chapter, hadronic mechanics permits the first and only known, exact and invariant representation of all characteristics of the neutron at the non-relativistic and relativistic levels. Note that the neutron can also be viewed as a compressed neutroid.

As indicated in Sect. 8.2.1, the above equation is grossly insufficient for the proposed neutron structure. Consequently, we assume that the neutron structure is characterized by additional contact, non-potential interactions that, to be outside the class of unitary equivalence of quantum mechanics, are characterized by a *non-unitary transformation* of the quantum unit 1 into the isounit $UU^\dagger = 1^* = \exp(V_h/V_c)$, with Santillian $S = \exp(—V_h/V_c) \approx 1 — (V_h/V_c)$, where V_h is the Hulten potential $W e^{-br}/(1 — e^{-br})$ and b is the inverse of the charge radius of the neutron.

We now insert the Santillian in between the Hamiltonian and the state of the reduced form for the hydrogen atom $[(1/2\mu)d/dr(r^2d/dr) — e^2/r]S|\psi^*(r)\rangle = [(1/2\mu)d/dr(r^2d/dr) — e^2/r](1 — V_h/V_c)|\psi^*(r)\rangle \approx [(1/2\mu)d/dr(r^2d/dr) — e^2/r — V_h]|\psi^*(r)\rangle = E|\psi^*(r)\rangle$ where, this time, E is the binding energy caused by the assumed non-potential interactions.

Recall that the Hulten potential behaves like the Coulomb potential at small distances, by therefore absorbing the latter due to its strength. The preceding equation is then well approximated by the hadronic reduced form $[(1/2\mu)d/dr(r^2d/dr) — V_h]|\psi^*(r)\rangle = E|\psi^*(r)\rangle$ which was solved in all necessary details in paper [3-5] under the subsidiary condition for the size and mean life of the neutron, resulting in the spectrum $E = K(1/n — n)^2$, where K is an isorenormalization constant (Eq. 5.1.20, p. 837 of memoir [3-5]), whose sole admissible value is that for $n = 1$, $E = 0$, as expected.

The important conclusion is therefore that, as expected, Santillian interactions have no role in the binding energies of the hadronic bound state. Consequently, binding energies are solely characterized by Hamiltonian interactions.

This important property has been confirmed by all applications of hadronic mechanics to bound states of extended particles with potential and non-potential interactions, such as electron valence bonds [4-9], stable nuclei [8-28]-[8-30], structure models of elementary particles with detectable constituents [8-31] and other models. In all cases, the deep entanglement of the wave packet of particles in single (triplet) coupling resulted to be strongly attractive (repulsive).

Since the electromagnetic binding energy of the neutron can be computed, in first approximation, via standard methods, let us recall that the binding energy of the Hydrogen atom is given by $E_H = (\alpha\hbar c/2)/R_B = 13.6$ eV, where $\alpha = 1/137$ is the fine structure constant and $R_B = 10^{-8}$ cm is Bohr's radius. By recalling that the radius of the electron orbit inside the proton is 10^{-13} cm $\approx 10^{-5}R_B$, the binding energy of the isoneutron is here predicted to be $E_{be} = 10^5 \times 13.6$ eV = 1360000 eV = 1.36 MeV.

Consequently, the rest energy of the neutron is predicted to be $E_n = E_p + E_e — E_{be} + K_e = 939.565$ MeV. By recalling that $E_p = 938.272$ MeV and $E_e = 0.511$ MeV, the kinetic energy of the electron is predicted to be $K_e = 2.172$ MeV with total energy $E^{tot} = 2.683$ MeV.

As indicated in Sect. 8.2.1, the origin of the missing energy of 0.782 MeV for the *mass excess* of the neutron over the sum of the masses of the electron and the proton, plus the *kinetic excess* of 1.390 MeV caused by the Coulomb binding energy, is by far the most mysterious and intriguing open problem in the neutron structure that will be addressed in detail in Sects. 8.2.8 and 8.2.9.

8.2.5. Representation of the neutron spin

The second technical problem for the proposed structure model of the neutron was the representation of the neutron spin according to Rutherford's original conception of $p^+ + e^- \rightarrow n$, that is, in the absence of the neutrino. In turn, such a problem was reduced to identifying the spectrum of eigenvalues of the angular momentum L_e of the electron when orbiting in two dimensions around the proton, as shown in Figure 8-2. According to quantum mechanics, the sole admissible eigenvalues for the angular momentum are given by $L_e = 0, 1, 2, ...$, resulting in the inability to represent the spin 1/2 of the neutron. This occurrence suggested W. Pauli's [8-3] hypothesis of the emission of a massless and chargeless particle with spin 1/2 called the *neutrino* by E. Fermi, $p^+ + e^- \rightarrow n + \nu$.

With due respect to the historical value of the Pauli-Fermi hypothesis, in Santilli's view, quantum mechanics is inapplicable to the dynamics of the electron compressed inside the hyperdense proton due to the emergence of non-linear, non-local and non-potential thus non-Hamiltonian interactions whose study stimulated the construction of the novel isomathematics (Chapter 3).

The value of the angular momentum of the electron inside the proton in a two-dimensional orbit can be derived as follows (See Ref. [8-14], Sect. 2.3.2 for a detailed presentation).

The transition from motion in a vacuum to motion within a physical medium can be rep- represented via the non-unitary transformation of quantum mechanical treatments $UU^\dagger = 1^* = \text{Diag.}(1/r^2, 1/r^2) = (1/r^2)\,\text{Dig.}(1.1)$, where the last identity results from the circle as being the most stable orbit, and r^2 represents the deviation from the orbit in the vacuum caused by resistive forces (technically done via isorenormalizations caused by the Santillian).

The application of the above non-unitary transformation to every aspect of the rotational symmetry O(3) (including numbers, functional analysis, differential calculus, etc.) yields the iso-rotational symmetry $O^*(3)$ [7-7], whose eigenvalues are $L^* = L_e/r^2$.

The condition that the angular momentum of the electron be equal to the spin of the proton to avoid extreme resistive forces yields the value $r = 2$ for which $L^* = J_p = 1/2$, under which the total angular momentum of the electron is null $J^{tot} = J_e — L_e = 0$, and the spin of the neutron coincides with that of the proton $J_n = J_p — J_e + L_e = J_p = 1/2$.

The first numerically exact and time-invariant representation of Rutherford's synthesis of the neutron from the proton and the electron without the conjecture of the neutrino was proposed by R. M. Santilli in the 1990 paper [8-8] *without* the indication that the treatment required no neutrino in respect for W. Pauli and E. Fermi. The expectation was evidently that the absence of the neutrino would be obvious following an inspection of paper [8-8] due to the lack of energy to produce the neutrino, let alone the impossibility of admitting its spin 1/2 for motion within a hyperdense medium.

The existence or nonexistence of the neutrino is deeply related to the used symmetries and relativities. In the event we assume the validity for the neutron synthesis of Poincare´s symmetry and related special relativity, the hypothesis of the neutrino appears to be unavoidable, as correctly predicted by W. Pauli and E. Fermi (Sect. 8-2.1).

However, Poincare´s symmetry and special relativities can be solely defined for isolated points in space-time, in which case the conservation of the total angular momentum $J_{tot} = J_p + J_e = 0$ of the Poincare´s symmetry requires the emission of a particle with spin 1/2 due to the prohibition by quantum mechanics for the angular momentum of a point-like particle in vacuum $J^{orb} = 1$, for which we would have $J_n = J_p + J_e^{spin} — J_e^{orb} = 1$.

By contrast, the representation via the iso-Poincare´s symmetry and related special isorelativity (Chapter 6) of the extended wave packets of the electron and the proton under their total mutual penetration and ensuing contact non-Hamiltonian interactions require, for stability, that the orbital angular momentum of the electron be equal to the proton spin, $J_e{*}^{orb} = J_p{*}^{spin} = ½$ resulting in the correct value of the spin of the neutron $J_n = J_p + J_{e*}^{spin} — J_{e*}^{orb} = ½$.

Additionally, linear momentum and angular momentum must be individually conserved for the Poincare' symmetry and special relativity while for their isotopic completion linear momentum can be transformed into angular momentum and vice-versa, by therefore prohibiting the application of the Pauli-Fermi hypothesis. The above important features of hadronic isosymmetries and isorelativities are clearly verified in nature because during its spontaneous decay the neutron emits an electron with the *linear kinetic energy* of 0.782 MeV that was originally stored inside the neutron in its *rotational form*.

Needless to say, the absence of the neutrino creates the problem of the representation of the data of the so-called "neutrino experiments" via alternative hypotheses which is treated in Sections 8.2.8 and 8.2.9.

8.2.6. Representation of the neutron magnetic moment

The third technical problem in the quantitative representation of the neutron synthesis from the Hydrogen is the numerical representation of the magnetic moment of the neutron $\mu_n = —1.913\ \mu_N$ (where μ_N is the nuclear magnetron) from the magnetic moment of the proton and of the electron, whose anomalous character is twofold. Firstly, the magnetic moment of the neutron is opposite that of the proton, and secondly, the magnetic moment of the neutron is expected to be null due to its null charge.

The technical difficulty here considered is compounded by the fact that, despite its small mass, the magnetic moment of the electron has the value $\mu^{spin} = —9.284764 \times 10^{-24}$ J/T $= —1838.2851\ \mu_N$, which is 961 times *bigger* than the neutron magnetic moment.

The significant difference between the electron and nuclear magnetic moments fueled the century-old, widely accepted view that *electrons cannot be nuclear constituents.*

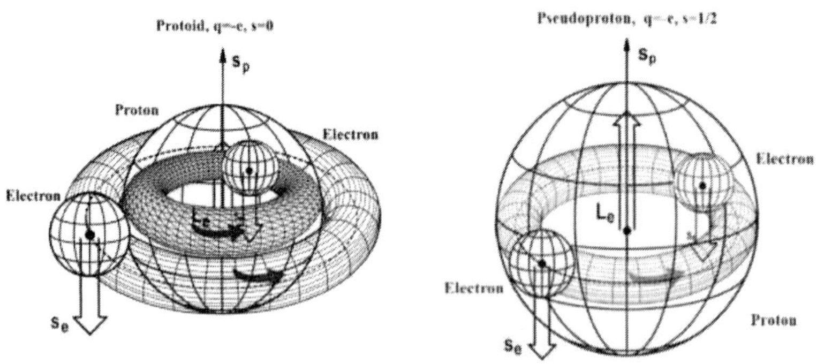

Figure 8-3. On the left, we illustrate the prediction of the protoid $\tilde{p}^-$ as a hadronic bound state of one electron, and this time, a neutron with the mass and mean life close to those of the neutron, negative charge, and spin-zero [8-34]. On the right, we illustrate the prediction of the pseudo-proton as a hadronic bound state

of an electron pair in singlet coupling totally compressed inside the proton, with mass and mean life close to those of the neutron, negative charge and spin 1. Note that the latter is a compressed form of the former, as was the case for neutroids and neutrons [8-34].

By contrast, Santilli argued in his1990 paper [8-8] that the *orbital* magnetic moment of the electron inside the spinning proton, μ^{orb} is of the same order of magnitude but *opposite* that due to its spin $\mu^{orb} \approx -\mu_e$, therefore voiding the general belief of the lack of electrons in nuclear structures. In fact, calculations first done in Santilli's 1990 paper [8-8] yielded the value $\mu^{orb} = +1833.5801\ \mu_N$ with the resulting numerically exact and time-invariant representation of the anomalous magnetic moment of the neutron as a hadronic bound state of the permanently stable proton and electron $\mu_n = \mu_p + \mu^{spin} + \mu^{orb} = 2.792\ \mu_N - 1838.285\ \mu_N + 1833.5801\ \mu_N = -1.913\ \mu_N$, which is at the foundation of the new applications of hadronic mechanics in nuclear physics, such as new means of recycling radioactive nuclear waste (Sect. 8.6) and new clean nuclear fusions (Sect. 8.7).

It should be noted that the assumption of the above orbital contribution of the iso-electron not only allows a representation of the numeric value of the anomalous magnetic moment of the neutron but also of its negative value.

8.2.7. Relativistic representation of the neutron synthesis

The numerically exact and time invariant representation of all characteristics of the neutron as a hadronic bound state of a proton and an electron via relativistic hadronic mechanics (rhm) was first reached in Santilli's 1995 paper [7-12] (and updated in the recent paper [8-14]) via the assumption that both the proton and the electron are mutated due to their deep mutual entanglement, resulting in Santilli's structure model of the neutron $n = (p^{*+}\uparrow, e^{*-}\downarrow)_{rhm}$ [8-22]-[8-26].

The study required the prior isotopy of each mathematical and physical part of relativistic quantum mechanics based on the isounit $1^* = 1/S = \text{Diag.}(1/b_1^2, 1/b_2^2, 1/b_3^2, 1/b_4^2) = \text{Diag.}(n_1^2, n_2^2, n_3^2, n_4^2)$, $b^2 = 1/n^2$, $\mu = 1, 2, 3, 4$, where the exponential term $\exp\Gamma$ is embedded in the dimensionless characteristic $b = 1/n$-quantities by therefore implying their highly non-linear, non-local and non-potential character.

More specifically, the relativistic treatment of the neutron structure required the iso-Lorentz $SO^*(3.1)$, iso-Poincare′ $P^*(3.1)$ and isospinorial $\mathcal{P}^*(3.1)$ symmetry (Chapter 7) for the indicated isounit and related Santillian with ensuing isotopies of special relativity and Dirac's equation. Regrettably, we cannot review these studies and must limit ourselves to the review of the main results of papers [7-12] [8-14] by referring interested readers to the quoted literature for all details.

Via the use of the indicated methods, paper [7-12] achieved a relativistic fit of the numeric values of the spin and magnetic moments of the neutron with the values $b_1 = b_2 = b_3 = 1.189$, $b_1^2 = b_2^2 = b_3^2 = 1.415$ Additionally, paper [7-12] identified the first known value of the *density* of the neutron $b_4 = 1/n_4 = 1.62$, $b_4^2 = 1/n_4^2 = 1.415$, a value in remarkable agreement with other densities of hadronic media studied in the next chapter, under which *the tangential speed of the electron's orbit inside the proton is superluminal* with $C = c\ b_4 = c/n_4 = 1.62$ c.

Similarly, the assumption of the neutron binding energy $E_{be} = 2.172$ MeV and of the total energy of the isoelectron $E^{tot} = 2.583$ MeV (Sect. 8.2.3), plus the use of the maximal causal speed $V_{max} = c\ b_4/b_3$, yields the density $1/n_4 = b_4 = 1.912$, $1/n_4^2 = b_4^2 = 3.659$ with the electron tangential speed $C = c\ b_4/b_3 = 1.607$ c.

It should be indicated that the superluminal speed of the electron when totally compressed inside the proton *is not* in conflict with the widely assumed belief that c is the maximal causal speed since, as stated by Albert Einstein various times, c is the maximal causal speed for *point-like* particles moving *in vacuum.*

By contrast, when we have *extended* particles moving within a *physical medium,* the ensuing contact non-Hamiltonian interactions can accelerate particles at arbitrary speeds since contact Santillian interactions occur without the use of energy (Sect. 8-2.3) as predicted by the 1982 paper [3-24] at the foundation of Santilli isorelativity for extended particles within physical media [3-25] (Chapter 7).

In other terms, the main open scientific issue is the *peripheral, tangential speed of the proton* when admitted with its extended size of 0.87×10^{-13} cm under the value 1/2 of its spin since the tangential speed of the relatively much lighter electron totally immersed in its interior is a mere consequence by keeping in mind, that again on scientific grounds, *special relativity is inapplicable for rotating bodies at large* evidently because of the inherent centripetal forces that violate the basic inertial assumption of said relativity.

In closing, we would like to indicate that the neutroid (Figure 8-1) is a clear intermediate state between the Hydrogen atom and the neutron (Figure 8-2) because the latter can be considered the compression of the former in the cores of stars.

8.2.8. Neutrino or etherino?

In this and the following section, we outline the status of our knowledge at this writing (February 14, 2025) on the most mysterious and intriguing open problem in nuclear physics, which is given by the missing energy in the neutron synthesis from the proton and the electron in the core of stars, $p^+ + e^- \rightarrow n + \nu$ (Sections 8.1-II and 8.2.4) that, for the tabulated $E_e = 0.511$ MeV, $E_p = 938.272$ MeV and $E_n = 939.565$ MeV, is given by 0.782 MeV $= E_n - (E_e + E_p)$.

Evidently, said missing energy casts shadows on the existence of the neutrino as a physical particle since 0.782 MeV is missing for the neutron mass alone, while the energy needed for the creation of the neutrino is nowhere available.

Said missing mass casts additional shadows on the conjecture that the non-detectable quarks are physical particles since the permanently stable protons and electrons cannot credibly be converted into the hypothetical quarks and at the time of the neutron decay, quarks cannot be credibly converted into permanently stable protons and electrons.

In this section, we outline the status of the studies on *Santilli etherino* [8-29] as the only known alternative to resolving the problem of missing energy while representing experimental data on the neutron decay.

In the next section, we outline the most important function of Santilli etherino, that of achieving the first known representation of the gravitational stability of the Sun [8-14]. The study of the inconsistencies of quark conjectures for the structure of the neutron and hadrons at large is left for brevity to interested colleagues [8-28] [8-47] [8-48].

For the intent of preserving the validity of special relativity and relativistic quantum mechanics in nuclear physics, it is generally conjectured that the missing energy of 0.782 MeV is provided for the neutron synthesis by the kinetic energy of the electron. However, this conjecture has been dismissed in serious scientific circles because stars synthesize approximately 10^38 neutrons *per second.* In the event the missing energy for the neutron synthesis is provided by the thermal energy in the star, the star would lose about 10^{38}MeV *per second*, by therefore cooling down and never producing light.

The cross-section of electron-proton scattering at 1 MeV relative energy is essentially null, in which case no fusion of the electron and proton into the neutron is credibly possible.

The immense pressure in the core of stars has no representation at all by special relativity and relativistic quantum mechanics, but it does have a quantitative representation in isorelativity (Chapter 7) and hadronic mechanics (Chapter 5) via the Santillian representing extended particles/wave packets with ensuing contact non-potential interactions. Under these conditions, the isostandard deviation of the linear momentum of the electron is essentially null, $\Delta p \approx 0$ (Chapter 6), with the consequential impossibility for the electron to carry a kinetic energy of about 1 MeV.

The complexity and implications of the problem considered are further illustrated by the following implications of the 0.782 MeV missing energy:

The first shadows on the conjecture of the neutrino as a physical particle were created by the century-old view that the Pauli-Fermi massless neutrinos could traverse entire astrophysical bodies without collisions. Additional shadows were created by the standard model, firstly, with the conjectured existence of various new neutrinos without clear physical differences from the Pauli-Fermi neutrino and, secondly, the additional conjecture that all hypothetical neutrinos have a finite rest mass.

As reported in nuclear physics textbooks, the energy carried by the electron in nuclear beta decays follows a bell-shaped curve with a maximum value of approximately 1 MeV. However, in view of the strongly attractive Coulomb interactions between the nucleus and the electron, the energy carried by the electron in nuclear beta decays is expected to depend on the direction of emission, with maximal value for radial emission and minimal value for tangential emission (see Sect. 2 and Fig. 1 of [8-29]).

Additional shadows on the neutrino as a physical particle are created by the impossibility of representing its spin 1/2 whenever the proton is represented as an extended particle in view of the constrained orbital motion of the electron compressed inside the proton, which hadronic model allows the exact representation of all characteristics of the neutron, including its spin (Sect. 8.2.5) and anomalous magnetic moment (Sect. 8.2.6).

Neutrinos have never been detected experimentally since all claimed evidence on the existence of neutrinos is indirect, that is, based on the detection of particles *predicted* by the neutrino hypothesis. The problem is that other models equally make the same predictions without the neutrino conjecture. Additional shadows are cast by the excessive paucity of the so-called "neutrino detections" (see, e.g., [8-49]), compared to the extremely large number of predicted neutrino events.

The flux of solar neutrinos reaching Earth is of the order of trillions of neutrinos per square centimeter per second. In comparison, their experimental detection has been extremely rare, prompting the Italian Government to terminate the neutrino tests at the Italian Gran Sasso Laboratory in 2006.

It should be noted that the neutron missing energy is not unique for the neutron synthesis because it also holds for the exact representation of all characteristics of unstable particles as hadronic bound states of lighter physical particles produced free in the spontaneous decays [8-28]. The latter problem requires the identification of the insufficiencies of quark conjectures, which is here left for brevity to interested colleagues (for details, see [8-48] [8-49]).

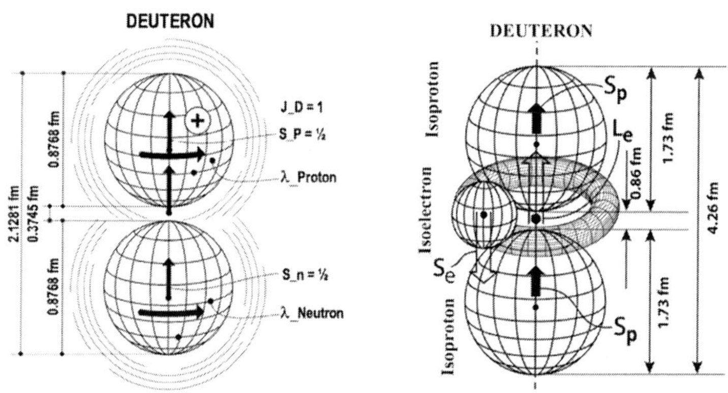

Figure 8-4. In the left, we illustrate the Deuteron as a proton-neutron bound state and, on the right, the Deuteron as a proton-isoneutron bound state according to hadronic mechanics.

In order to initiate quantitative studies of alternatives to the neutrino conjecture, Santilli proposed in the 2007 paper [8-29] the hypothesis that the missing energy in the synthesis of the neutron from protons and electrons originates from the ether as a universal substratum with extremely high energy density and said energy is transferred from the ether to the neutron by a longitudinal impulse called *etherino* (meaning in "little ether" in Italian) and represented with the symbol a (from the Latin aether) with mass, charge and spin 0 and a minimum of 0.782MeV energy, according to the hadronic synthesis $p^+ + a + e^- \rightarrow n$, where one should note the presence of the etherino in the l.h.s. of the synthesis as a necessary condition to carry the missing energy. By comparison, the quantum synthesis $p^+ + e^- \rightarrow n + \nu$ places the neutrino in the r.h.s. because it is needed to conserve the angular momentum for quantum mechanics without consideration for the missing energy.

The hadronic decay of the neutron (when isolated) is given by $n \rightarrow p + e$, whose electron carries the original missing energy, which is therefore released into our world. In this way, the neutron synthesis with the acquisition of 0.782 MeV from the ether as a universal substratum and consequential "mass excess," and its release into our world via the spontaneous decay $n \rightarrow p + e$. This model constitutes the best known mechanism for Hoyle's continuous creation of matter in the universe, also known as steady state cosmology [8-49].

We should note again from Chapter 7 that the decay $n \rightarrow p++e$ violates the conservation of the energy for point-like particles represented via quantum mechanics. However, the same decay verifies the special isorelativity in view of the admission by the Poincaré-Santilli isosymmetry P*(3.1) the conversion of the orbital into kinetic energy (see Fig. 3 of [8-29]).

Additionally, explicit calculations indicate that the immense energy needed for a supernova explosion cannot be explained via the sole use of conventional nuclear fusions, also in view of the fact that supernova explosions occur at the end of the life of stars, thus suggesting again the possible existence of a mechanism extracting energy from the ether and transforming it into a form existing in our spacetime.

It is important to point out that the notion of ether as a universal substratum appears to be necessary not only for neutron synthesis but also for the characterization and propagation of electromagnetic waves and all elementary particles.

We should also note that being a longitudinal impulse, the etherino is predicted to propagate at a speed much bigger than the speed of the electromagnetic waves as predicted by the Lorentz-Santilli isosymmetry (Sect. 7.3.3), thus being significant for inter- stellar communications alongside the interstellar communications that are conceivably possible for the hadronic entanglement (Chapter 6).

Another supporting feature is that the etherino hypothesis eliminates the implausible belief that massive particles carrying energy in our spacetime can traverse large astrophysical bodies without collisions since the corresponding etherino event would occur via the propagation of longitudinal impulses through the ether as a universal substratum rather than through hyperdense matter.

We encourage interested colleagues to verify that, to our best knowledge, all predictions based on the neutrino hypothesis $p^+ + e^- \rightarrow n + \nu$ are equally made by the etherino hypothesis $p^+ + a + e^- \rightarrow n$ when treated in first approximation as a quantum mechanical reaction among point-like particles, with the understanding that the admission of the ex- tended character of wave packets implies a necessary revision of the quantum scattering theory [4-5].

We finally indicate that Santilli ended his paper [8-29] on the etherino hypothesis with proposals for the experimental resolution of the existence or lack thereof of neutrinos, including the proposals for the direct measurements of the missing energy (Sect. 9 and Fig. 6 of [8-29]) and of the electron energy in neutron decays (Sect. 9 and Fig. 7 of [8-29]).

8.2.9. The gravitational stability of the Sun

The Sun releases into light the energy of [8-50] $\Delta E^{Sun} = 2.3 \; 10^{38}$ MeV/seconds, which corresponds to about $4.3 \; 10^{6}$ tons/sec. Since, in a Gregorian year, there are 10^{7} seconds,

the loss of mass by the Sun per year ΔM^{Sun} due to light emission is given by $\Delta M^{Sun} = 10^{23}$ metric tons per year.

The above loss of mass by the Sun is of such a magnitude to cause an increase in planetary orbits that should be detectable by contemporary astrophysical laboratories contrary to centuries of measurements on the stability of planetary orbits.

A medium-sized star, such as our Sun, synthesizes about 10^{38} neutrinos *per second* [8- 51], which requires a total energy of about $\Delta E^{star} = 10^{38}$ MeV per second.

The etherino hypothesis was formulated on the grounds that the energy needed for the neutron syntheses by the Sun is essentially equal to the Sun's loss of energy into light. Consequently, the assumption that the ether provides the missing energy for neutron synthesis as a universal substratum permits the first and only known quantitative representation of the stability of the Sun's gravitational field.

8.2.10. Reduction of matter in the universe to protons and electrons

In the 2024 paper [8-14], Santilli presents the ultimate implication of his studies in nuclear physics, which is the reduction of all matter in the universe to the permanently stable protons and electrons.

According to centuries of scientific knowledge, stars initiate their lives as aggregates of Hydrogen that increase by accretion during their travel in hydrogen-rich interstellar spaces. When the pressures and temperatures in the star reach certain values, a star initiates the synthesis of neutrons from protons and electrons. Neutrons then bond with protons by forming Deuterons and initiating the production of light via the fusion of two Deuterons into Helium and via subsequent fusions.

In view of the above evidence, nuclei were originally conceived to be bound states of protons and electrons [8-52] [8-53]. However, this original conception had to be abandoned with the advent of quantum mechanics because of the experimentally unverified *assumption* that the validity of Heisenberg's uncertainty principle for *point-like* particles in vacuum also holds for *extended* protons and neutrons under strong nuclear forces, resulting in the *theoretical assumption* that electrons cannot be constituents of nuclear structures on the grounds of:

8.2.10A. Excessive electron speed. By recalling the value of the electron mass $m_e = 0.511$ MeV $= 9.1 \times 10^{-31}$ Kg and the nuclear radius $R = 10^{-14}$ m, *Heisenberg's uncertainty principle* $\Delta r \Delta p = |\langle\psi|[r, p]|\psi\rangle| \geq h/2 = 5.26548578 \times 10^{-34}$ JHz^{-1} implies that the electron has the superluminal velocity $v \geq 5.79 \times 10^{10}$ m/s.

8.2.10B. Excessive electron energy. Under the validity of Heisenberg's principle, a nuclear electron would have the standard deviations of the momentum $\Delta p = 1.05\ 10^{20}$ kg m/s, with corresponding energy $E = 19.5$ MeV, contrary to the evidence that electrons emitted in Beta decays have a maximum energy of 3 MeV.

8.2.10C. Excessive magnetic moment. The magnetic moment of the electron expressed in nuclear magnetron μ_N has the value μ_e = —9.284764×10^{-24} J/T = —1838.285 μ_N, which is 2321 times the magnetic moment of the deuteron μ_D = 0.857 μ_N.

As a result of half a century of preparatory mathematical, theoretical and experimental studies reviewed in the preceding sections and chapters, Santilli [8-14] resolved the above objections by allowing indeed electrons to be physical constituents of nuclear structures. In particular, objections 8.2.10A and 8.2.10B were resolved by the deep mutual entanglement of the weave packets of the proton and of the electron represented via a very big isounit $1^* = 1/S \gg 1$ representing nuclear internal pressures, hence, with a very small Santillian $S \ll 1$, the Heisenberg- Santilli isouncertainty principle $\Delta r^* \Delta p^* = (1/2)\, |\langle \psi^*| \star [r^{*\hat{}}, p^*]^* \star |\psi^*\rangle = (1/2)S \ll 1$ for which an isolated electron in a nuclear structure is kept mostly at rest by the extreme pressures caused by the strong nuclear interactions. Objection 8.2.10C was resolved by the fact that the orbital motion of the electron compressed inside the proton, being equally large but opposite that of the electron magnetic moment, reduced the latter value down to nuclear standards (Sect. 8.2.6).

Following the above preparatory studies, Santilli achieved the reduction of all matter in the universe to protons and electrons via his reduction of the neutron to a hadronic bound state of a proton and an electron reported in the preceding sections of this chapter. The reduction of all matter in the universe to protons and electrons was then completed by the proof that the *permanently stable protons and electrons* cannot be credibly transformed into hypothetical undetectable particles such as the quarks.

Intriguingly, the main principle that sustained Santilli in his decades of studies in the field is that the electron in *spontaneous* nuclear beta decays cannot be synthesized from other particles and must, therefore, be a nuclear constituent because the permanent stability of the electron is at the foundation of the very existence of the universe.

8.3. The syntheses of Santilli's protoids and pseudo-protons

8.3.1. Foreword

As it is well known, nuclear physics has been unable to achieve controlled nuclear fusion primarily due to the Coulomb barrier, namely, the natural repulsion between nuclei resulting from their positive charge, which becomes extremely large at a mutual distance of the order of 10^{-13} cm, necessary to activate strong nuclear interactions.

Following the representation of the synthesis of the neutron from the Hydrogen, in an attempt to resolve the Coulomb barrier, Santilli initiated studies on the synthesis of [8-30]-[8-35], as hadronic bound states of natural (positively charged) nuclei and (negatively charged) electrons, which study, predictably, had to initiate at the level of strongly interacting particles.

8.3.2. The synthesis of protoids

An inspection of the plasma in the interior of stars (when admitted as being ultimately composed of the originating, permanently stable electrons and protons) reveals that the hadronic mathematical and physical methods for the synthesis of neutroids and neutrons predict additional hadronic bound states, the first being the *protoid* $p^{-} = (e^{-}\downarrow, n^{0}\uparrow)_{hm}$ which is a hadronic bound state of an electron, and this time a neutron in singlet coupling (left of Figure 8-3) with mass and mean life of the order of those of the neutron elementary negative charge and spin $J_{p^-} = 0$. Besides the extremely strong Coulomb attraction between protons and electrons, the protoid is created by the strongly attractive non-Hamiltonian forces of deep entanglements of wave packets originating from the neutroid (Sect. 8.2.2).

8.3.3. The synthesis of pseudo-protons

The prediction of the protoid implies the prediction of the *pseudo-proton* p^{*-} [8-34], which is a hadronic bound state of an electron totally compressed within the neutron, resulting in an unstable state with the mass and mean life close to those of the neutron, negative charge and spin ½.

Finally, hadronic mechanics predicts a bound state of an electron pair in singlet coupling totally compressed, this time, inside the proton, $p^{*-} = ([(e^{-}\downarrow, e^{-}\uparrow), p^{+}]_{hm}$ (Figure 8-3) resulting in a third strongly interacting particle with mass and mean life close to those of the neutron, negative charge and spin 1, the latter value being due to the fact that the total spin of the electron pair is null $J_{2e} = 0$ but, for stability, its angular momentum inside the proton must be equal in value and direction to the proton spin $L_{2e} = 1/2$, resulting in the indicated total value of the spin $J = 1$.

Note that the plasma in a star is expected to contain a significant presence of the hadronic pair $(e^{-}\downarrow, e^{-}\uparrow)_{hm}$ called *isoelectronia* (first represented with an attraction by hadronic chemistry [4-9]) evidently because it originates from the ionization of the primordial Hydrogen gas into two protons and an isoelectrium.

The representation of all characteristics of the protoids and pseudo-protons at the non-relativistic and relativistic levels are essentially those for the neutroid and neutrons of Sect. 8.2 [8-34] [8-35], and their study is left to the interested reader.

8.4. Representation of the Deuteron and pseudo-Deuteron

8.4.1. Foreword

Since the Deuteron is considered to be a primary building block for all nuclei [8-36], Santilli has conducted comprehensive studies [8-37] - [8-42] for the numerically exact and time invariant representation of all experimental data of the Deuteron at both non-relativistic and relativistic levels. The studies were conducted via structure models of the Deuteron of increasing complexity, from the century-old proton-neutron bound state to advanced models with Bohm's hidden variables.

8.4.2. The Deuteron as a proton-neutron hadronic- bound state

In the author's view, the biggest insufficiency of quantum mechanics in nuclear physics is its inability to represent the spin 1 of the smallest nucleus, the Deuteron, which spin, according to experimental measurements, is the spin of the *Deuteron ground state* that, by definition, requires that all orbital contributions are null, while quantum mechanics has represented the spin of the Deuteron for about one century via orbital contributions.

This historical occurrence is due to the fact that the sole possible quantum mechanical ground state of two particles with spin 1/2 is the singlet, in which case the spin of the Deuteron would be $J_D = 0$. Consequently, for the intent of maintaining the validity of quantum mechanics for the Deuteron structure, nuclear physicists have represented for about one century the spin $J_d = 1$ of the Deuteron via a collection of orbital contributions in violation of the conditions for its experimental detection.

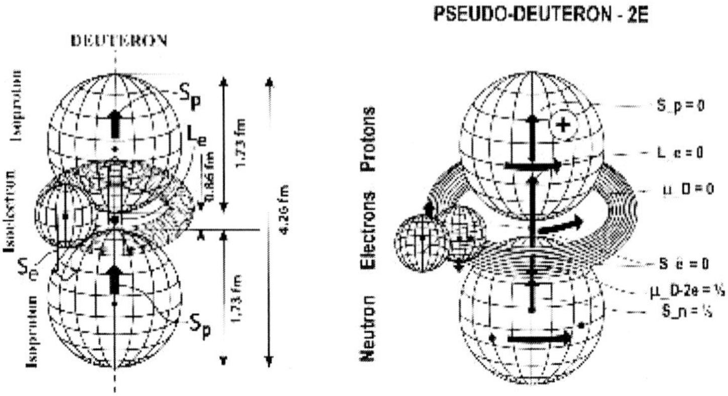

Figure 8-5. We illustrate, on the left, the positively charged Deuteron as a proton-isoneutron bound state, and on the right, the negatively charged pseudo-Deuteron-2 according to hadronic mechanics.

In passing to heavier nuclei, the Helium-4 is known to be a bound state of two Deuterons in singlet coupling. The above quantum mechanical representation of the spin $J_D = 1$ gives the impression of a correct representation of the spin of the Helium, $J_{He} = 0$, while in reality, it implies the implausible, permanent existence inside the permanently stable Helium-4 of excited orbital states of protons and neutrons. Hence, the insufficiency of quantum mechanics to represent nuclear spins at large emerges quite strongly.

Santilli has first shown in paper [8-31] that the historical conception of the Deuteron as a bound state of a proton and a neutron does indeed allow an exact representation of the Deuteron data, provided that the representation is done via hadronic mechanics, namely, by representing the charge distributions of protons and neutrons as being ex- tended, thus deformable in conditions of partial mutual penetration/entanglement with ensuing contact/non-potential interaction represented by a Santillian with the now familiar realization $S = \Pi_{k=1,2}\text{Diag.}(1/n_1^2, 1/n_2^2, 1/n_3^2, 1/n_4^2) \times \exp—\Gamma \ll 1$.

In fact, the above assumptions imply the resolution of the inconsistency for the Deuteron spin because hadronic mechanics admits the bond of two spin 1/2 particles along their spin axis, called *axial triplet coupling* (see the left image of Figure 8-4), for which the total spin is correctly given by $J_D = 1$. Additionally, the axial triplet coupling is the Deuteron ground state because all excited states imply the disintegration of the Deuteron due to mutual distances becoming bigger than the range of strong interactions.

Recall from Chapter 1 that the quantum mechanical representation of the Deuteron magnetic moment is about 3% in *excess* of the experimental data, with much bigger deviations for heavier nuclei. During a 1994 visit to the Joint Institute for Nuclear Research, Santilli [8-37] showed that the indicated quantum insufficiency is due to the point-like approximation of protons and neutrons because the representation of nucleons as being extended, thus deformable with their actual size of 0.87 fm, does indeed allow a numerically exact representation of the Deuteron magnetic moment which is invariant under the iso-Galileo symmetry and relativity.

Santilli reinspected and upgraded his representation of the Deuteron magnetic moment a number of times. The last of the deuteron magnetic moment in (3+1)-dimensions is provided in Sect. 5.4 of the 2024 paper [8-31] yielding the following numeric values of the characteristic quantities $n_1 = n_2 = 1.1432$, $n_3 = 0.6215$, $n_4 = 0.605$ (Eqs. (91) of [31]) according to which

HELIUM

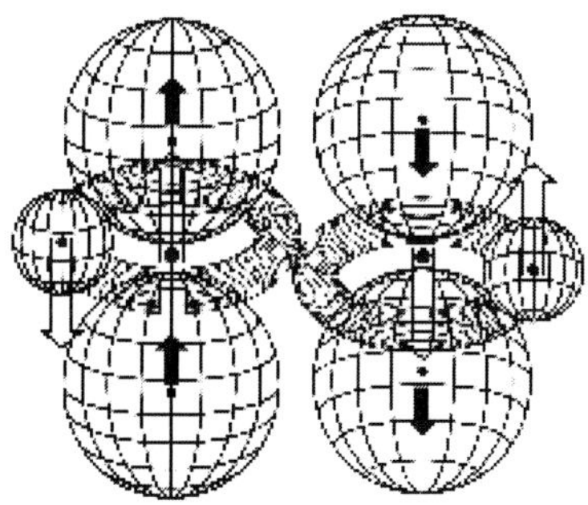

Figure 8-6. A view of the Helium-4 according to hadronic mechanics.

8.4.2A. The exact representation occurs with approximately 1% prolate deformation of the charge distributions of the proton and the neutron, caused by their strong nuclear interactions.

8.4.2B. The density of the deuteron turns out to be the same as that of the neutron (Sect. 8.3.7), as well as that of high energy scattering.

8-4-2C. In accordance with Iso-Axioms I of special isorelativity (Sect. 7.3.6), particles can propagate within the Deuteron with the superluminal speed $C = c/n_4 = c/0.605 = 1.652 \times c$, e.g., for tangential speeds for which special relativity is structurally inapplicable due to centrifugal accelerations.

The representation of the mass, charge radius and stability of the Deuteron was done in Sect. 5.7 of paper [8-31] via the same structure equation (Sect. 8.2,4) and the representation of the strong nuclear interactions via the Hulten potential $V_h = -e^{-br}/(1 - e^{-br})$.

Besides the exact representation of the considered Deuteron data, an important result of the studies has been the confirmation that the non-potential interactions represented by the Santillian carry no energy because the binding energy of the Schrödinger-Santilli isoequation with the indicated Santillian solely admits the null value.

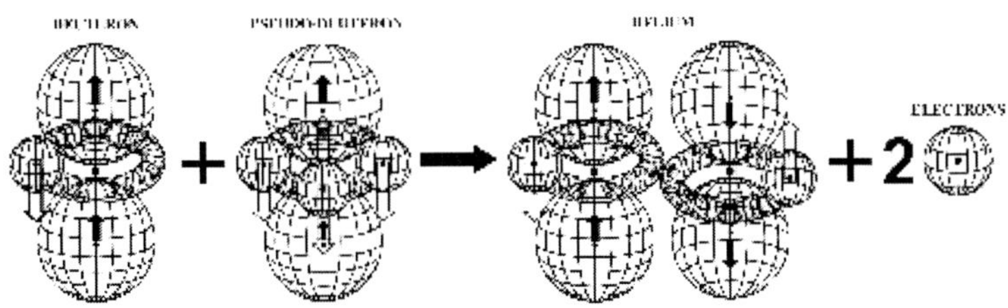

Figure 8-7. A view of the HyperFusion according to hadronic mechanics.

Furthermore, the finite spectrum of the Hulten potential is restricted to the ground state as the sole admissible state by confirming that the Deuteron does not admit strongly interacting excited states, evidently because all excited states would imply mutual distances between the proton and the neutron bigger than the range of strong interactions.

8.4.3. The Deuteron as a proton-isoneutron hadronic- bound state

In Santilli's view, all his preceding studies on the Deuteron were incomplete because they assumed the neutron to be a permanently stable particle while, in reality, the neutron is naturally unstable with a mean life of about 15 minutes.

Following several attempts, Santilli finally achieved in the 2024 paper [8-32] a quantitative representation of nuclear stability by showing that, when a member of the Deuteron or heavier nuclei, the permanently stable electron decouples from the neutron structure under the Coulomb attraction by the nearby proton (left view of Figure 8-5).

Paper [8-32] shows that the exact representation of the Deuteron data of the preceding paper [8-31] persists under the decoupling of the neutron electron and that such a decoupling allows the first known explicit and concrete realization of nuclear exchange forces, which are evidently given by the exchange of said electron in between the two protons.

Paper [8-32] completes the studies in nuclear stability by showing that nuclear forces absorb the Coulomb repulsion between nuclear protons and remain so strong to assure nuclear stability.

This important property was proved as follows. Consider the quantum mechanical Schrö dinger equation of the He(4, 2, 0, u_{he}) as a two-body bound state of two Deuterons in singlet coupling $[-(1/m_D)\Delta_r + V_c]|\psi\rangle = E|\psi\rangle$ showing with the "+" sign the extremely repulsive Coulomb potential in between the two Deuterons. In order to map the above quantum equation into its hadronic counterpart, introduce the following non-unitary transformation $UU^\dagger = 1/S = Exp(+v_h/V_c) \approx 1 + V_h/V_c$ (Eqs. (63) of [8-32]) where V_h is the Hulten potential used earlier in this section. Then, the above equation is turned into the Schrödinger-Santilli isoequation $[-(1/mD)\hat{\Delta} + \hat{V}_c]S|\hat{\psi}\rangle = E|\hat{\psi}\rangle$ that, following simple passages, is reduced to $[-(1/m_D)\Delta_r + V_c - V_h]|\psi\rangle = E|\psi\rangle$. But, at very small distances, V_h behaves like the Coulomb potential while being much stronger than the latter, by therefore absorbing the Coulomb potential with a mere re-definition of its constant and reducing the Helium-4 structure equation to the form $[-(1/m_D)\hat{\Delta} + V_c - V_h]|\psi\rangle = E|\psi\rangle$ which constitutes the first and only known representation of the permanent stability of the Helium-4 despite the extremely repulsive Coulomb force between its two protons with the astronomical value (for particle standards) of 230 Newton.

In the concluding remarks of both papers [8-31] [8-32], Santilli stresses the prior need for the above results of the hadronic entanglement of particles (Sect. 6.3), the hadronic generalization of Bell's inequalities (Sect. 6.4) and the ensuing generalization of Heisenberg's uncertainty principle (Sect. 6.5).

8.4.4. The Deuteron as an isoproton-isoneutron hadronic bound state

In Section 8.2.10, we have reviewed the reduction of all matter in the universe to protons and electrons. Such a reduction is technically based on the notion of isoparticles of hadronic mechanics, which is too complex for a review in this book.

The French Duke Louis de Broglie and David Bohm (Sect. 3.5) pioneered the fact that, under sufficiently small mutual distances, the numeric values of the characteristics of a particle depend on the interactions with the surrounding particles. The de Broglie-Bohm non-local interactions were assumed at the foundation of hadronic mechanics (Chapter 3) and represented via the Santillian S, as these interactions are of a contact, thus zero-range type, not derivable from a quantum potential.

Despite such a conceptually evident scenario, the technical realization of the indicated reduction is too complex for its being reported in this book.

We merely mention that, in the transition from the neutron to the Deuteron and then to heavier nuclei, protons and electrons acquire a corresponding, progressive complexity of intrinsic characteristics, which have been represented in paper [4-14] via the notions of isoprotons and isoneutrons as isounitary isoirreducible isorepresentations of the isospinorial covering of the Poincaré-Santilli isosymmetry $P^*(3.1)$ with a progressively increasing complexity and corresponding progressive chain of relativities indicated in Chapter 7.

The result important for this section is that, despite these complexities, the structure model of the Deuteron as a hadronic bound state of two isoprotons and one intermediate isoelectron (right view of Figure 5) does indeed represent all experimental data of the Deuteron [8-14] with sufficient degrees of freedom, including explicit and concrete realizations of Bohm's hidden variables, to be usable as a building block for all stable nuclei [8-36].

8.4.5. Representation of pseudo-nuclei and pseudo-Deuterons

As indicated in Section 8.4.1, the biggest obstacle against the industrial achievement of controlled nuclear fusion is the repulsive force between two natural nuclei due to their positive charges, called the Coulomb barrier, which acquires simply astronomical values (for particle standard) of the order of hundreds of Newtons for the fusion of two Deuterons into the Helium, and of the order of thousands of Newtons for the fusion of heavier nuclei. It is evident that, under such powerful repulsive forces, engineering means to achieve controllable and sustainable nuclear fusion are believed to be impossible for our current technologies, thus requiring new vistas.

In order to resolve the Coulomb barrier, Santilli proposed in the 2022 paper [8-33] the existence of a bond between electrons e^- and natural nuclei N(A, Z, J), called *pseudo-nuclei*, according to the reaction $n\ e^- + N(A, Z, J) \rightarrow N^*(A^*, Z^*, J^*)$, where $Z^* = Z — n$ and $J^* = J — J_e$ is the singlet coupling of the original spin J and the total electron spin J_e.

The existence of pseudo-nuclei N^* was based on the extremely strong Coulomb attraction at short distances between electrons and natural nuclei, which is represented by the Hamiltonian H, as well as the equally strong contact interactions due to deep mutual entanglements of electrons-nuclei wave packets at short distances represented by the Santillian S in the Heisenberg-Santilli time evolution of an observables $idA/dt = [A, H]^* = A \star H — H \star A = ASH — HSA$ (Chapter 3). Needless to say, pseudo-nuclei are unstable, but their half-life is predicted by hadronic mechanics to be of the order of seconds, thus being amply sufficient for industrial applications (see Sect. 8.8.4 for the experimental verification of their existence).

Evidently, for nuclei with big atomic number A, the charge of pseudo-nuclei is expected to remain positive, $Z — n > 0$, due to the engineering impossibility of bonding to a natural nucleus a large number of electrons. In this case, the Coulomb barrier between heavy natural nuclei and pseudo-nuclei persists.

By contrast, for *light natural elements,* the charge of pseudo-nuclei can be negative, $Z^* = Z — n < 0$, in which case the Coulomb barrier is not only resolved but turned into an extremely strong attraction.

In order to illustrate this possibility, Santilli introduced in paper [8-33] the hadronic bound state of an electron pair in singlet coupling and the Deuteron, resulting in the pseudo-Deuteron D^* also called D_{2e}, according to the reaction $(e^-\uparrow, e^-\downarrow)_{hm} + D(2, 1, 1) = D_{2e}(2, —1, 1)$ (right of Figure 8-5), called pseudo-Deuteron-2e, where one should note the acquired *negative* charge, whose bound state is impossible for quantum mechanics (particularly in view of Heisenberg's uncertainty principle) but fully possible for hadronic mechanics.

Under the above premises, hadronic mechanics predicts the fusion between the pseudo-Deuteron-2e and a natural Deuteron into the Helium $D_{2e}(1,2,1) + D(2,3,1) = He(2,4,0) + 2e^-$ [8-33], which verifies the following features (see Figure 8-7):

8.4.5A. Said fusion is *spontaneous* within a Hydrogen plasma because it occurs naturally due to the strongly attractive Coulomb force between the natural Deuteron and the pseudo-Deuteron-2e in addition to the strongly attractive nuclear forces in singlet coupling which occur at contact;

8.4.5B. Said fusion occurs under the equally spontaneous antiparallel alignment of spins due to their magnetic moments and,

8.4.5C. Said fusion releases 23.847 MeV per reaction without the emission of harmful radiations and the release of radioactive waste.

Hadronic mechanics predicts the existence of the pseudo-Deuteron D_{2e} in the core of stars, by therefore facilitating nuclear fusions in a star, as well as in the plasma of a DC arc submerged within a Deuterium gas since its ionization generates a plasma containing electrons, valence electron pairs and Deuterons.

The characteristics of the pseudo-Deuteron-2e are predicted to be the following (Sect. 3.3 of [9-33]):

8.4.5-I: Mass given by 1876.812 MeV from the mass of the Deuteron 1875.612 MeV plus the mass of the two electrons 1.220 MeV less the Coulomb binding energy of 20 MeV.

8.4.5-II: Spin the same as that of the natural Deuteron, J = 1, because the electron pair has null spin and null angular momentum since the pseudo-Deuteron-2e is considered in its ground state

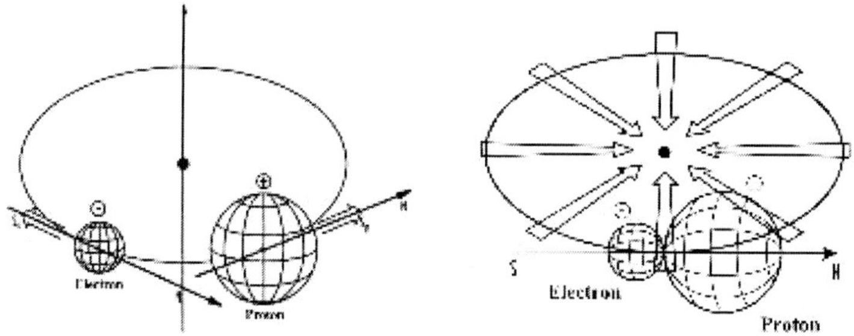

Figure 8-8. On the left, we illustrate the ionization of the Hydrogen by the activation of a DC arc with the ensuing alignment of the electron and proton along a local magnetic line. On the right, we illustrate the compression by the DC arc during its disconnect of the electron-proton pair in all radial directions toward the arc symmetry axis, resulting in neutron synthesis.

8.4.5-III: Magnetic moment predicted to be 3.666 μ_N from the magnetic moment of the orbital motion of the electron compressed inside the proton in the neutron structure.

8.4.5-IV: Mean life, which is assumed in paper [8-33] to be 1 sec because sufficient to produce all needed nuclear fusions, while calculation suggests a mean life of 10 sec.

8.4.5-V: Spontaneous decay $D_{2e}(—1,2,1) \rightarrow D(1,2,1)+2e^-$.

The structure equation of the pseudo-Deuteron-2e has been studied in great detail in Sect. 3.3 of [9-33] via methods similar to those used for the structure of the neutron and of the Deuteron.

8.5. Experimental confirmation of neutroids and neutrons

8.5.1. Foreword

Following, and only following the theoretical representation of the synthesis of neutroids and neutrons from the Hydrogen in the core of stars [8-8]-[8-14], Santilli conducted tests from 2006 to 2016 on their laboratory syntheses and possible industrial applications under the following main guidelines:

8.5.1A. Use of specially designed *hadronic reactors,* i.e., reactors designed via the laws of hadronic mechanics, which essentially consist of chambers certified to hold a commercially available Hydrogen gas at a pre-set pressure while being traversed by a DC arc between submerged electrodes.

In essence, Don Borghi [8-7] and his associates used the electric arc to maintain the hydrogen gas ionized inside the klystron and thought to use microwaves to synthesize neutrons. By contrast, theoretical studies on neutron synthesis, confirmed by the tests considered here, indicate that microwaves are unable to compress the electron inside a proton.

The latter basic requirement can instead be met via a suitable DC arc between submersed electrodes because, when inspected with oscilloscopes, DC arcs are continuously connected and disconnected due to electrode consumption. As illustrated in Figure 8-6, during its activation, the DC arc ionizes the Hydrogen and aligns electrons and protons along the local magnetic force with their subsequent coupling in singlet (because the magnetic moment of the electron has a direction opposite that of its spin). During its disconnection, the DC arc compresses the so- coupled electron and proton in all radial directions toward the arc symmetry axis, rearranges the coupling into a singlet and activates (under sufficient power) the strongly attractive interactions due to deep entanglements that, combined with the extremely big Coulomb attraction between electrons and protons, compresses the electron-proton pair into the neutron.

8-5-1B. Use the *minimal* DC power needed to synthesize neutrons with kinetic energy of 100 eV or less, hereinafter referred to as thermal energy, in order to achieve the following novel applications.

A number of neutron sources are commercially available today, but they all produce high-energy neutrons, which have consequential limited industrial applications due to special shielding and other requirements. A primary objective of Santilli's research was the production of a neutron flux synthesized on demand from hydrogen with minimal possible kinetic energy, thereby maximizing potential industrial and medical applications.

8.5.1C. Reactors with DC power less than 0.782 MeV = 1.24×10^{-13} J per synthesis are primarily intended for the laboratory synthesis of thermal neutroids, while reactors with DC power bigger than the indicated threshold (e.g., DC arcs with 10^{-9} J sufficient to produce a neutron flux of 100 cps) are intended for the synthesis of thermal neutrons.

As indicated in Sect. 8.3, the most mysterious and, therefore, intriguing open problem in the synthesis of the neutron from a proton and an electron in the core of stars is the origin of the missing energy of 0.782 MeV. Since the missing energy cannot be consistently provided by the stars to prevent its quenching due to the synthesis in a star of about 10^{38} neutrons *per second* and for other reasons, Santilli [8-29] submitted the hypothesis that the ether provides the missing energy as a universal substratum via a mechanism essentially implying the continuous creation of matter in the universe.

Calculations via the data of Don Borghi's experiment indicate insufficient DC energy to synthesize neutrons, therefore confirming Don Borghi's expectation that he had synthesized neutroids.

8.5.2. Experimental confirmation of Don Borghi's neutroid

We now review the first detection of neutroids reported in the 2007 paper [9-15]. All tests were initiated on July 28, 2006, at 2 p.m, at the laboratory of the Institute for Basic Research (IBR) in Palm Harbor, Florida, with the participation of the IBR technicians Terry Allen, John T. Judy, Michael Rodriguez, Jim Alban and Ray Jones

The first tests were done with a vertical, cylindrical, hadronic reactor (Figure 8-7) of about 6" outside diameter and 12" height made of commercially available, transparent, PolyVinyl Chloride (PVC) containing a commercial Hydrogen gas at about 15 psi, the cylinder housing a pair of tungsten electrodes of 0.250" diameters and 1" length fastened to conducting rods protruding through seals out of the top and bottom of the reactor to power and control the DC arc powered by a Miller Sincrowave 300 with 5 KV A delivering about 300 A at 20 V.

The PVC was selected to be transparent so as to allow a visual detection of the arc. The radiation detectors used in tests [8-15] are:

1) A photon-neutron detector model PM1703GN manufactured by Polimaster, Inc., in Russia, with sonic and vibration alarms with the photon channel activated by CsI and the neutron channel activated by LiI;
2) A photon-neutron detector SAM 935 manufactured by Berkeley Nucleonics, Inc., in California, with the photon channel activated by NaI and the neutron channel activated by He — 3 equipped with a sonic alarm;
3) A BF — 3 activated neutron detector model 12 — 4 manufactured by Ludlum Measurements, Inc., New Jersey;

4) An alpha, beta, gamma and X-ray detector model 907 — palmRAD manufactured by Berkeley Nucleonics, Inc.; and
5) Various materials are suitable for nuclear transmutations.

The main results of tests [8-15] with the indicated hadronic reactors are as follows.

A) Systematically, no particle detection occurred immediately following the activation of the arc, to such an extent that, at times, detections were obtained by impacting the reactor with a rubber mallet. This systematic event confirms that, under the indicated power of 5 kV A, the arc primarily synthesizes particles that are mostly confined to the electrode gap and then releases them with an energy of 100 eV or less, primarily due to collisions.

B) Systematically, minutes following the activation of the arc, Santilli detected particles originating from objects surrounding the reactor rather than from the reactor itself, including the casing of the detectors.

C) A comparative study of the detections by the used instruments established that the reactor synthesized a neutral particle with essentially the same mass, charge radius, and main life as the neutron but with spin zero, thus being Don Borghi's neutroids.

Let us denote nuclei with the standard symbol N(A, Z, J, u), where A denotes the atomic number, Z the proton number, J the spin and u the mass, and let us denote neutroids with the symbol $\tilde{n}(1, 0, 0, 1.007)$, where 1.007 u is the mass of the neutroid here assumed to be equal to that of the proton. Analysis of the data indicated that the reactor produced neutroids that were adsorbed by stable nuclei in surrounding equipment by creating the new class of *nucleoids* $\tilde{N}(A+1, Z, J, U+1.007)$ that, being unstable, generally decay via electron capture (EC) or other processes [8-29] by reproducing the original nucleus plus the emission of neutrons detected by the instruments $\tilde{N}(A+1, Z, J, u+u_{\tilde{n}}) + EC \rightarrow N(A, Z, J - 1/2, u) + n(1, 0, 1/2, 1.008)$. The above mechanism also explains that no stable substance exposed to neutrons in ten years of tests showed any residual radioactivity; that is, following the absorption of neutrons and the subsequent expulsion of detected neutrons, the substances returned to their original stable status.

The following episode provided the final confirmation of the synthesis of neutroids. On July 30, 2006, Santilli conducted a test with the Polimaster, which was located next to the reactor. Following three minutes with no detection, Santilli left the laboratory with the Polimaster in his pocket and, on his way home, stopped at the Walgreens Pharmacy in Tarpon Springs, Florida, located about 12 minutes driving distance from the laboratory, to purchase a toothpaste but, while being in line for payment, the Polimaster in his pocket entered into violent sonic and vibration alarms such to cause fig-11 Santilli's rapid exit from the pharmacy so as not to trigger the arrival of security personnel.

Following this episode, Santilli returned to the laboratory, repeated identically the preceding tests and left the laboratory with the Polimaster in his pocket to go home via a different route. Following fifteen minutes of driving, the Polimaster in Santilli's pocket entered again into a violent sonic and vibrational alarm.

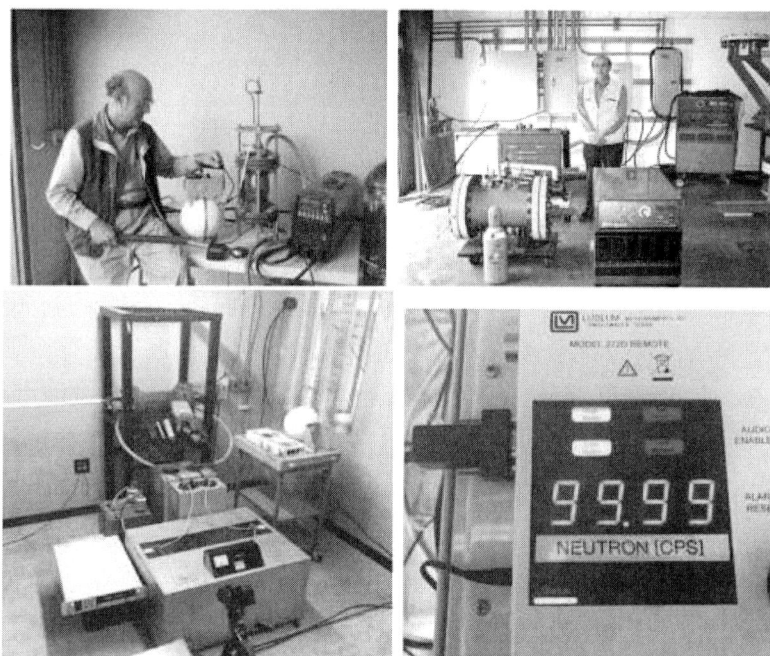

Figure 8-9. In this figure, starting from the top left, we illustrate the small hadronic reactor with 5 kV A used to detect neutroids, the bigger reactor with 50 kV A used to detect neutrons, an alternative version of the latter, and one of the neutron counts during the latter tests.

In the years following the 2006 tests [8-15], Santilli and his associates tested a variety of different hadronic reactors described in subsequent papers [8-16]-[8-21] (Figure 7), which all confirmed the detection of neutroids under the indicated data.

8.5.3. Experimental confirmation of the neutron synthesis

Besides the hadronic reactor with 3 KV A reviewed in the preceding section, Santilli [8- 15] constructed in 2006 a second reactor (top right of Figure 8-7) comprising a pressure-resistant, horizontal, metal vessel schedule 40 of 12" outside diameter and 24" length containing a Hydrogen gas at 100 psi traversed by a DC electric arc powered by a 50 kV A AC-DC converter Miller Dimension 1000 delivering about 1000 A at 40 V, said arc occur- ring through a gap of about 1/8" between 1/2" cylindrical tungsten electrodes controlled on the outside, axially located in the interior of the reactor and remotely operated for safety. After flushing out the internal atmosphere and filling the reactor with hydrogen at 100 psi, the test of the second reactor, surrounded by all available detectors, was conducted on December 20, 2006.

In short, at the instant of the remote activation of the 50 KV AC-DC power, all neutron detectors triggered violent sonic and vibration alarms, immediately halting the test and evacuating the building for the rest of the day.

The same test was repeated on January 15, 2007, with the same results that were confirmed in subsequent years with additional hadronic reactors and neutron detectors [8-16] [8-21], thereby establishing the laboratory synthesis of the neutron from hydrogen beyond scientific doubt.

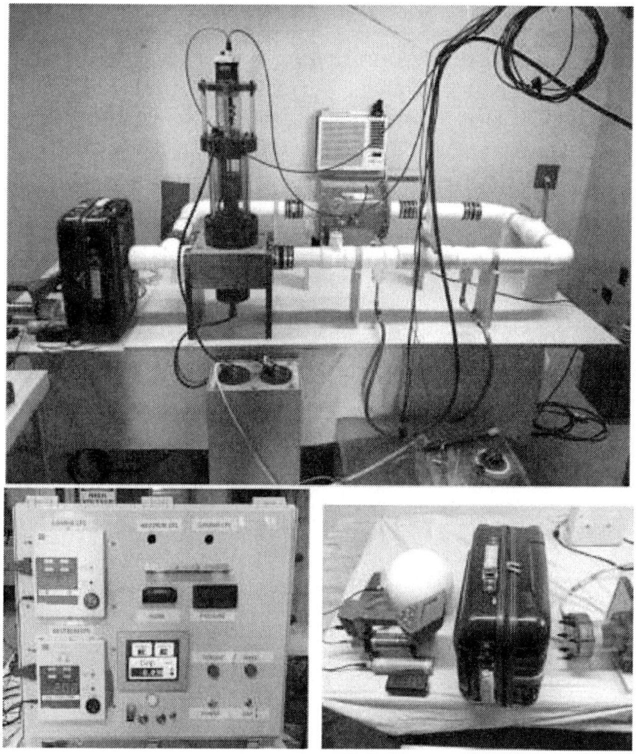

Figure 8-10. In these pictures, we illustrate the Directional Neutron Source (DNS) produced and sold by the US publicly traded company Thunder Energies Corporation (now Hadronic Technologies Corporation) from 2015 to 2019.

8.5.4. The Directional Neutron Source (DNS) and its industrial applications

The biggest threat facing our societies nowadays is the smuggling of the components of a nuclear bomb and its detonation on our soil. Our intelligence has done an incredible job in the background to prevent this from happening, despite the lack of scanners in our airports and ports, which would provide clear detection of fissile material that may be concealed in baggage. In fact, nuclear fuels such as Uranium-233, Uranium-235 and Plutonium-239 are metals that, as such, cannot be clearly distinguished from other metals via X-rays, microwaves or other currently available scanning equipment (see Ref. [8-43] for details). The most effective way to detect smuggled fissile material is by irradiating baggage or grounds with a flux of very low energy neutrons (in the range of a few eV) because, in this case, some of the fissile nuclei disintegrate by releasing a shower of heavy radiation which is clearly identified by an available neutron, gamma and other detectors. A number of neutron sources are available these days, but they all release very high- energy neutrons (in the range of several MeV) that cannot be used in public areas.

Figure 8-11. We illustrate the equipment used for the separation of neutrons and pseudo-protons, the former being absorbed by the casing of the generator and the latter being released into its exhaust.

Following decades of mathematical, theoretical and experimental studies on the synthesis of the neutron from the proton and the electron, in his capacity as Chief Scientist of the U. S. publicly traded company Thunder Energies Corporation (now Hadronic Technologies Corporation), Santilli successfully achieved the industrial synthesis on demand of very low energy neutrons (of the order of eV) from a commercially available hydrogen gas that can be used as duly shielded scanning equipment to detect fissile material that can be concealed in baggage or soils [8-44].

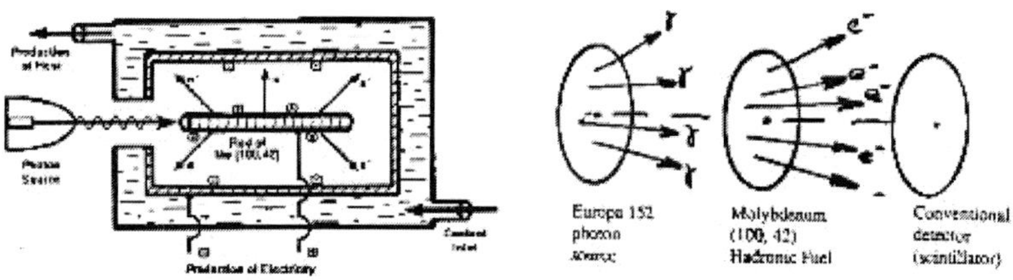

Figure 8-12. On the left of this figure, we reproduce Santilli's [5-55] original 1994 drawing on the proposed stimulated beta decay, and on the right, we illustrate its experimental verification by N. Tsagas and his associates [8-59] as well as the low cost of such an important, yet widely ignored test (Sect. 8.7.2).

The *Directional Neutron Source* (DNS) model TEC-DNS- 3B-5 with remote operating panel (Figure 8-10), manufactured and sold by the U. S. publicly traded company *Thunder Energies Corporation* (now Hadronic Technologies Corporation) from 2015 to 2019, produces on demand a flux of neutrons, synthesized by a DC arc submerged in Hydrogen, in the preferred direction with energy ranging from 10eV to 100eV and Neutron Counts per Second (CPS) ranging from 5 CPS to 90 CPS. The directional character of the neutron flux is achieved via the flow of the hydrogen gas through the DC arc. The desired neutron energy is achieved by varying the DC power from 5 kW to 12 kW. The desired neutron CPS is achieved by varying the Hydrogen pressure from 5 psi to 20 psi. TEC-DNS-3B-5 is primarily intended for:

1) Scanning baggage or soil for possible concealed fissile material; 2) Use by academic or corporate physics laboratories; 3) Scanning large welding in naval constructions and other uses. Due to the general human presence in its vicinity, TEC-DNS-3B-5 has been designed and constructed to have the "minimal" possible neutron energy and counts.
2) However, TEC-DNS with a smaller or bigger neutron energy range and smaller or bigger neutron CPS can be built on request.
3) It should be indicated that the TEC DNS produces a directional flux primarily containing neutrons but also containing neutroids, protoids and pseudo-protons in about 10% of the treated volume.

8.5.5. Experimental confirmation of protoids and pseudo-protons

In preparation of the experimental verification or denial of the synthesis of pseudo-protons (Section 8.3.3 and Figure 8-3), Santilli [8-45] studied their radiations by all new particles studied in this section, including the irradiation of the pure Silver-47 isotope with pseudo- protons Ag(47, 107, 1/2) + p^{*-}(—1, 1, 1/2) → N*(46, 108, 1) →Pd(46, 108, 0) + γ(0, 0, 1) which produces the stable isotope Pd(46, 108, 0). Consequently, the identification in their- radiated Ag-47 of synthesized Pd(46, 108, 0) over impurities of Pd(46, 110, 0) (sufficiently over the background) would provide clear experimental evidence on the existence of the Santilli pseudo-proton. Note that isotope N*(46, 108, 1) is not tabulated, and therefore, it is a pseudo-nuclide.

The Directional Neutron Source (DNS) of the preceding section could not be used for pseudo-proton activation because of its primary production of neutron flux, besides having minimal power for human safety. Also, the dual irradiation by neutrons and Santilli pseudo-protons produced by the DNS would damage the clarity of the test.

Consequently, Santilli constructed the equipment of Figure 8-11, releasing a flux pre-dominantly composed by pseudo-protons without detectable neutrons and consisting of a commercially available General Motor 14 kW electric generator model 01705, 3 Liter displacement, 4 cylinders operating on natural gas, hereon referred to as the GMEG. The standard ignition of this GMEG was upgraded with a commercially available MSD ignition system, specifically the MSD 8 plus 7805 with Coil, Blaster HVC-2, and U-Cor, which delivers sparks with 45 kV at 235 J.

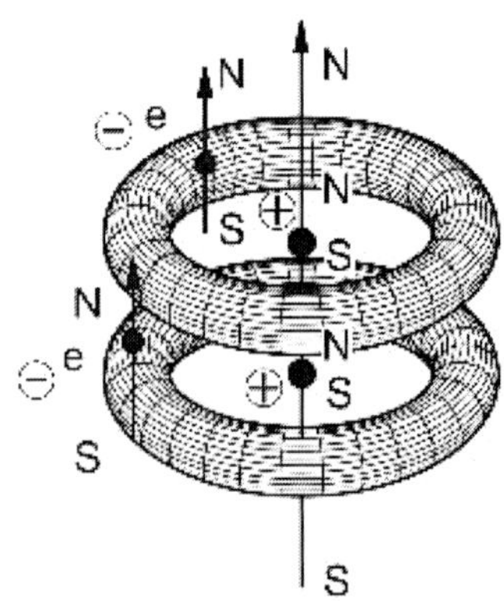

Figure 8-13. We illustrate the new chemical species of magnecules [8-62]-[8-71], which have been developed to achieve a controlled and sustainable exposure of nuclei outside their electron clouds (Hadronic Law 8.7.4-I).

This upgraded GMEG was used in the tests as follows: the GMEG was placed and operated for safety in the interior of a metal container; the air intake was sealed with a metal cup and related fastener; in lieu of natural gas, the generator was fed with Hydrogen at 30 psi yielding a total Hydrogen compression pressure at the time of ignition estimated to be of 150 psi; the generator was started via the use of its standard cranking system operated by a battery connected to a battery charger; the exhaust terminal was covered with a metal mesh incorporating the selected samples of a pure Silver isotope (Figure 8-11, lower left view).

Direct measurements via a Berkeley nucleonics neutron detector model SAM 934 placed in the exhaust (Figure 8-11, lower right view) showed no detectable neutron in the exhaust. Consequently, neutral particles synthesized by the engine sparks, including neutrons, were absorbed by the engine casing, while all charged particles, including Santilli pseudo-protons, were instead released through the engine exhaust and irradiated the pure samples of Silver placed at its terminal.

Figure 8-14. A view of the ICNF hadronic reactor for the HyperFusion of Carbon-12 and Deuterium-2 into Nitrogen-14 (Sections 8.7.6-I, 8.7.6-II and 8.8).

To initiate the experimental confirmation or denial of the existence of the pseudo-proton, the author selected the comparative measurements in parts per million (ppm) of Pd(46, 108, 0) over Pd(46, 110, 0) on samples of non-irradiated and irradiated samples of pure Silver. In early February

2017, two samples of pure Silver 2 cm square and 2 mm thick were purchased from ESPI Metals with certification of impurities. One sample was kept away from any irradiation, while the second was exposed to the exhaust of the GMEG for about 7 minutes of operation at the indicated conditions. Under the supervision of technician Michael Rodrigues, the samples were duly prepared and labeled and then shipped to the EAG Laboratory for comparative measurements. The EAG report dated March 7, 2017, EAG Job No C0HVP280, provided clear spectroscopic evidence of excess Pd(46, 108, 0) over impurities of Pd(46, 110, 0) by, therefore, providing clear evidence of the existence of Santilli pseudo-proton [8-45].

On September 2018, Santilli shipped two samples of non-irradiated and irradiated pure Silver to the Center for Plasma Mass Spectrometry-Core Research Facility, University of California, Davis, which released on October 10, 2018, the results of their comparative analysis by confirming the excess of Pd(46, 108, 0) over impurities of Pd(46, 110, 0) and therefore, confirming the existence of Santilli pseudo-proton [8-45].

An extensive study on the experimental detection of pseudo-protons has been conducted by V. DeHaan [8-20] and can be summarized as follows.

Pseudo-protons are negatively charged particles with a mass comparable to that of the Hydrogen atom or proton. It can be discriminated from an electron in a mass spectrometer due to the large mass difference and can be distinguished from a proton because of its reversed charge. In a Hydrogen discharge, hydride ions will also be produced. The mass and charge of the hydride ion are comparable to that of the pseudo-proton.

Hence, it is possible that in past experiments, the pseudo-proton was detected but interpreted as a hydride ion.

The mass difference between the hydride ion and the pseudo-proton is expected to be very small, of the order of 0.1%. Hence, when using a mass spectrometer to determine what kind of particle is detected, the accuracy should be high enough to measure this small difference. An additional test might be performed by measuring the magnetic moment of the detected negatively charged particle using an appropriate device, such as a Penning trap.

Mass spectrometers can be bought off the shelf, but one with an accuracy of 0.1% for a mass of approximately 1 amu is not readily available and must be developed. In such cases, it can be combined with measurements of magnetic moments.

8.5.6. The DeHaan pseudo-protonium

V. DeHaan noted [8-20] that the pseudo-proton can quickly capture a proton to form an atomic-type bound state that he called a *pseudo-protonium atom,* where the proton and pseudo-proton circle around each other like in a positronium.

Hence, one can imagine that in a strong electrical discharge in Hydrogen gas, the following reaction might occur: $H_2 \rightarrow H + H \rightarrow p^{*-} + p^{+} \rightarrow (p^{*-}, p^{+})_{hm}$.

Hence, in the event confirmed, pseudo-protons can form pseudo-protonium atoms during the discharge process. These atoms have energy levels determined by the energy levels of a two-body system. The reduced mass of this system is about 1836 times bigger than that of the hydrogen atom, resulting in energy levels in the keV range. These levels can be detected using X-ray spectrometry to study the Lyman transitions from levels $n > 1$ to $n = 1$ and the Balmer transitions from levels $n > 2$ to $n = 2$.

The intensity of the spectral lines depends on the number density of the pseudo-protonium atoms and hence will depend strongly on the discharge parameters and the pressure of the Hydrogen gas. Suppose these series can be found in the spectra of Hydrogen gas discharges with sufficient accuracy. In that case, the reduced mass of the pseudo-protonium atom and, hence, the mass of the pseudo-proton can be measured.

It should be noted that in some energy-dispersive X-ray spectroscopy experiments, peaks are recorded that are close to the estimated value, giving clues under what conditions pseudo-protonium might be formed.

8.6. The recycling of radioactive nuclear waste according to hadronic mechanics

8.6.1. The societal problem of recycling radioactive nuclear waste

Due to known public opposition, it appears that the recycling of radioactive nuclear waste [8-54] should be done by nuclear power plants in their facilities, hence, via their stimulated decays.

Unfortunately for our society, the reduction of the natural mean life of all unstable nuclei, including those of radioactive nuclear waste, is prohibited by special relativity (because it would violate the basic Poincare´s symmetry), and this occurrence explains the reason for the lack of any appreciable academic-governmental research in the field despite its need for mankind.

In this section, we review a number of stimulated decays of radioactive nuclear waste predicted by hadronic mechanics [8-55] [8-56] [8-57] [8-20] [8-31], which may reduce mean lives from thousands of years down to seconds, thus being usable by nuclear power plants in their facilities with ensuing production of new nuclear energy.

It should be recalled that individual scientists or corporations are prohibited from handling radioactive nuclear waste. Consequently, in this section, we review new forms of recycling nuclear waste that the appropriate governmental agency should test. We should finally indicate that in view of the very short half-lives of the new particles to be experimentally verified, compared to the extremely short half-lives (of the order of 10^{-30} s) of the particles detected at physics laboratory, in the preceding sections we assumed in the first approximation that neutroids, neutrons, protoids, pseudo-protons and other particles are stable, thus allowing the use of the Lie-isotopic branch of hadronic mechanics with basic reversible Heisenberg-Santilli isoequation for the time evolution of observables $idA/dt = [A, H]^* = A \star H — H \star A = ASH — HSA$, where H is the Hamiltonian representing potential interactions, and S is the Santillian representing the extended character of nucleons and their consequential contact non-potential interactions.

From this section on to the end of this chapter, we are studying strictly time- irreversible processes whose consistent treatment requires the Lie-admissible branch of hadronic mechanics with basic irreversible Heisenberg-Santilli genoequation for the time evolution of observables $idA/dt = (A, H) = A < H — H > A = ARS — HSA$, where R and S $\neq$ R are the forward and backward Santillians in the standard realization with $R = 1$ and $S = 1 — F/H$ given by $idA/dt = (A, H) = [A, H] + FA$, where F represents the historical Lagrange's and Hamilton's external terms originating the irreversibility of nature (Chapter 3) and related mathematics (Chapter 4), hadronic uncertainties (Chapter 6) and relativities (Chapter

Figure 8-15. A view of the ICNF hadronic reactor for the HyperFusions of Carbon and various gaseous feedstocks (Sects 8.7.6-I to 8.7.6-V).

Hadronic mechanics predicts three possible methods for recycling radioactive nuclear waste via their stimulated decay described in the following sections none of which is admitted by special relativity to prevent the violation of the Poincare' symmetry, but they are readily possible for its irreversible Lie-admissible covering. These methods can be combined with a variety of alternative methods available from the U. S. Patent Office which are too technical to be reviewed in this book. All methods here insured produce new clean nuclear energies described in the next section.

8.6.2. Recycling radioactive nuclear waste via neutron irradiation

This first method aims at the increase of the nuclear atomic number A while maintaining the protonic number Z via irradiation with very low energy neutrons such as those produced by the *Directional Neutron Source* of Sect. 8.5.4 and Figure 10.

We are here referring to nuclear transmutations of the type $N(A, Z, J) + n \rightarrow N^*(A + 1, Z, J + 1/2)$ where N* is not generally tabulated because it's an extremely unstable pseudo-nucleus which decay in lighter elements. As an illustration, consider the nuclear waste element given by the 55-Cesius-136.

Its stimulated transmutation via neutron irradiation is given by Cs(55 ,135, 7/2) + n → Cs*(55, 136, 5) + e⁻ → Ba(56,136, 0) + 2 e⁻ and does constitute a recycling of the original waste element because the mean life of the 55-Cs-136 is 13.16 days while the produced 56-Ba-156 is stable and the half-life of the pseudonucleus N* is so short not to be tabulated. Additionally, the rest energy of the Cs is 135.907306 while that of the Ba is 135.904570 resulting in the production of nuclear energy without harmful radiation for 0.04 u = 37.26 MeV per reaction.

This first recycling can be studied via nuclear transmutations of stable nuclear elements of the type Mo(42, 100, 0) + 2 n → Tc(42, 102, 1), which transforms the unstable 42-Mo-100 with a half-life of 10^{19} *years* into the isotope 42 -Mo -102 with a half-life of one minute.

Note that the proposed transmutations generally violate special relativity and quantum mechanics because of the possible lack of conventional conservation of nuclear spins. This problem does not exist for hadronic symmetries and relativities due to the possible interchange between linear and angular momentum.

As an illustration, the total spin of the original neutron irradiation in triplet coupling Cs(55, 135, 7/2) ↑ + n↓ is S = 3. This value can be preserved by hadronic mechanics in the final transmutation into Ba(56,136, 0) + 2 e- via the emission of an electron pair in singlet coupling 2e = (e⁻ ↑ , e⁻ ↓) and orbital momentum L = 3, or via an orbital angular momentum between 55-Ba-136 and 2 e⁻ of the detected type [8-106], or other forms of entanglements.

8.6.3. Recycling radioactive nuclear waste via stimulated neutron decay

This second method aims at maintaining the nuclear atomic number A while increasing the protonic number Z via the stimulated decay of peripheral neutrinos (Sect. 8.7.2) with transmutations of the type Cs(55, 135, 7/2) + γ_r → Ba(56, 135, 3/2) + e⁻. In this case, the half-life of 55-Cs-135 is $7{,}25\times10^{13}$ years while the produced 56-Ba-135 is stable.

8.6.4. Recycling radioactive nuclear waste via pseudo-proton irradiation

This third method aims at increasing both the atomic number A as well as the protonic number via irradiation within very low energy pseudo-protons (Sect.9.7) and nuclear transmutations of the type Cs(55, 135, 7/2) + p^- → Ba(56, 136, 3/2) + 2 e^- resulting again in the production of clean nuclear energy. A corresponding transmutation of stable natural elements suggested for study is given by Mo(42, 100, 0) + 2$\bar{p}$(—1, 1, 0) → Zr(40, 102, 0), in which case the 10^{19} years main-life of 42-Mo-100 is reduced to 2.9 seconds mean-life of 40 - Zr -102.

Note that the advantage of this third method over the preceding two methods is that low energy pseudo-protons are attracted by the nuclei of the radioactive waste, by therefore implying a bigger efficiency in the recycling. Note finally that pseudo-protons are predicted to be produced by the Directional Neutron Source jointly with the primary neutron production. As a result, this third method is essentially contained in the first method of irradiation via thermal neutrons.

In the author's view, the above application of pseudo-protons and protoids should not be dismissed on grounds of its incompatibility with the *antiprotons* allegedly produced at various particle laboratories in view of their unsettled character as true antiparticles [5-58]. It is hoped that particle physics laboratories may develop reactors that produce a beam of antiprotons or pseudo-protons to test the possible recycling of radioactive nuclear waste.

8.7. Clean nuclear energies according to hadronic mechanics

In our view, the search for controlled and sustained nuclear fusions based on quantum mechanics should evidently continue. Nevertheless, the lack of achievement in industrial applications, despite three-quarters of a century of attempts at the cost of trillions of dollars in public funds, suggests the need for fundamentally new, clean, controllable, and sustainable nuclear energy sources.

Following a lifetime of mathematical, physical, chemical, and experimental preparatory research reviewed in the preceding chapters, this section reviews the status of industrial development of essentially new, controllable, and sustainable nuclear energies predicted by hadronic mechanics to occur without the emission of harmful radiation and the release of radioactive waste. The new clean energies are collectively called *hadronic energies* [8-55] because they are based on the Einstein-Podolsky-Rosen completion of quantum into hadronic mechanics (Chapter 3).

8.7.1. Clean nuclear energies via stimulated beta decay

The hypothesis that the neutron is a hadronic bound state of a proton and an electron implies its possible stimulated decay when the neutron is a member of suitably selected nuclei, and it is irradiated with photons γ_r having the resonating frequency equal to the total energy of the iso-electron, e.g., Er * frequency of 3.129×10^{20} Hz or integer.

Additionally, said stimulated decays imply the possible production of controllable and sustainable nuclear energy without the emission of harmful radiation and the release of radioactive waste. This appears to be the case for the stimulated neutron decay first proposal by Santilli on page 340 of the 1994 paper [8-55] $\gamma_r + Mo(100, 42, 0) \rightarrow Tc(100, 43, 1) + \beta^- \rightarrow Ru(100, 44, 0) + 2$ β^- with the production of two different types of clean energies:

8.7.1A. Clean thermal energy, given by $\Delta E = E(100, 42, 0) - E(100, 44, 0) = 3.034$ MeV less the resonating photon energy of 1.293 MeV plus the energy 2.586 MeV of the two emitted electrons, resulting in the production of 4.327 MeV per reaction.

8.7.1B. Clean electric energy is given by the production of two easily trapped electrons for the production of a DC.

Figure 8-12 outlines on the left the original proposal on page 340 of [8-55] (see also book [9-13]) to illustrate the controllable and sustainable character of the new clean energy. It is set up consisting of a cylinder of Mo(100, 42, 0) irradiated from one side with resonating photons and surrounded first by a metal casing that, in turn, is surrounded by a heat exchange system. The metal casing captures the emitted electrons to produce DC electricity, while the heat exchanger captures the entire heat released by the transmutation. Controllability of the process is established by the halting of all energy releases at the disconnect of the resonating photon, while sustainability is established by the long duration of the Mo(100, 42, 0) bar.

A second illustration of the stimulated beta decay was provided on page 340 of [5-55] via the use of the stimulated transmutation γ_r+30-Zn-

70 $\rightarrow$ 31-Ga-70+β^-, which converts the stable isotope 30 – Zn - 70 into the unstable pseudo-nucleus 31-Ga-70 with decay 31-Ga-70 $\rightarrow$ 32-Ge-70 + β^- resulting in 32-Ge-70 which is a stable natural element. The initial element has an atomic weight of 69.925, while the final element has a smaller atomic weight of 69.924, resulting in the production of 931.494 MeV plus the energy of two beta decays.

This second example confirms that stimulated neutron decay transmutes a given element into another element with a smaller mass and different proton number, releasing a dual clean nuclear energy source because no massive particle escapes from the casing of Figure 9-12.

8.7.2. Tsagas confirmation of Santilli's stimulated beta decay

Santilli's 1994 predictions of stimulated beta decay [8-55] have been preliminarily confirmed by the Greek experimental team headed by N. F. Tsagas [8-59] via the setup on the right of Figure 8-12 illustrating the simplicity as well as low cost of the experimental verification that implies:

(A) The purchase of small samples of commercially available isotope Mo(100, 42, 0) and the radioisotope Europa-152 (emitting photons with E_r).

(B) The placement of the isotope Mo(100, 42, 0) and the radioisotope Europa-152 next to each other.

(C) Expose the pair of the isotope Mo(100, 42, 0) and radioisotope Europa-152 to low-cost detectors such as scintillators or other electron detectors.

Figure 8-16. A view of the ICNF hadronic reactor for the HyperFusions of Carbon-12 and Oxygen-16 into Silicon-28 (Sections 8.7.6-VI and 8.8).

Tsagas team provided the first experimental confirmation [8-59] of Santilli's stimulated beta decay [8-55] (see also review [8-24]) via the detection of electrons emitted by the pair Mo(100, 42, 0)/Europa-152 that could only be emitted by the indicated stimulated beta decay

Besides the production of clean thermal and electric nuclear energies, the stimulated beta decay is predicted to have, in due time and under proper funding, additional important applications, such as the recycling of radioactive nuclear waste (Sect. 8.7.1); the artificial transmutation of erroneous nuclei in a DNA into correct nuclei; the synthesis of rare elements from readily available elements, rather than their ongoing (very expensive and very polluting) chemical separation from rare earths.

8.7.3. Clean nuclear energies via pseudo-proton irradiations

Let us consider the negatively charged spin 1/2 pseudo-proton $p^{*-} = (n\uparrow, e^{-}\downarrow)_{hm} = [p^{-}, (e^{-}, e^{-})]_{hm}$ (Figure 8.3 and Ref. [8-19]) with evident instantaneous decay when a member of a nuclear structure $p^{*-} \rightarrow p^- + (e^-, e^-)_{hm}$. The irradiation of natural elements via pseudo-protons is predicted in Ref. [8-60] to produce the following transmutations $p^{*}_{-} + N(A, Z, J) \rightarrow N^{*}(A+1, Z-1, J+1/2)$ and therefore, change the atomic and proton numbers of natural elements.

Recall that Santilli's stimulated beta decay (Sect. 8.7.2) is only possible for a selected number of natural elements due to various conservation laws. By comparison, pseudo-proton transmutations are universal in the sense that they hold for all natural elements due to the very strong Coulomb attraction between pseudo-protons and natural nuclei with inevitable physical contact and activation of strongly attractive nuclear forces.

Needless to say, the transmuted pseudo-nucleus $N^{*}(A + 1, Z - 1, J + 1/2)$ is extremely unstable, thus explaining its lack of official admission by contemporary nuclear physical societies, with instantaneous decays that may indeed produce clean nuclear energies.

As an illustration, we consider the nuclear transmutation of 3-Li-7 via pseudo-proton irradiation $p^{*-}(1, -1, (1/2)_{\uparrow}, 1.008) + Li(7, 3, (3/2)_{\downarrow}, 7.016) \rightarrow Li^{*}(8, 3, 1/2, 8.0225) + \beta^- \rightarrow Be(8, 4, 0, 8.0225) + 2\ \beta^- \rightarrow 2\ He(4, 2, 0, 4.0026) + 2\ \beta^-$, where one should note [8-60]: the antiparallel spin polarizations of the pseudo-proton and the Li(7, 3, 3/2, 7.016) triggering the release of the first excess electron; the necessary intermediate production of the pseudo-nuclide $Li^{*}(8, 3, 1/2, 8.0225)$ that does not exist in the standard Table of Nuclides [8-105]; its instantaneous decay into the tabulated isotope Be(8, 4, 0, 8.0225) with the half-life of $8.19\ \ 10^{-17}$ s and its decay into two He(4, 2, 0, 4.0026) with the release of the clean nuclear energy of $2.887\ \ 10^{-12}$ J.

In case confirmed, the produced energy of $2.887\ \ 10^{-12}$ J per transmutation has industrial value because at the rate of 10^{16} stimulated transmutations per hour on a mole of Li-7 with 10^{23} nuclei and an efficiency rate of 10^{-7}, it would yield the clean nuclear energy of 1.7×10^{6} J/hour. Note that the above-stimulated transmutation would produce two high-energy electrons that can be collected to produce heat and electric energy.

For the study of additional nuclear transmutations via pseudo-proton irradiation of Pd-106 and Au-197, we refer the interested reader to Eqs. (10), (11) of [8-60].

Note finally that pseudo-protons can be produced via available technologies, such as the use of a conventional piston engine fueled by hydrogen, as originally done in Refs. [8-19]. Moreover, pseudo-protons can be produced via the same technology currently used to produce antiprotons at international physics laboratories because pseudo-protons have the same mass, spin and charge as antiprotons. Their sole difference is in the interpretation of pseudo-protons as ordinary particles, while antiprotons are interpreted as antiparticles [8-58].

8.7.4. Physical laws of controlled nuclear fusions according to hadronic mechanics

Following the representation of the neutron synthesis from protons and electrons and its use for the exact representation of nuclear data, the physical laws of new clean nuclear energies predicted by hadronic mechanics were presented for the first time in Santilli's 1998 monograph [8-61] and can be expressed as follows:

HADRONIC LAW 8.7.4-I: Hadronic fusion reactors should have means for the systematic and controlled exposure of nuclei out of their electronic clouds. In fact, in the absence of said exposure, no systematic and controlled nuclear fusions are possible due to the protection of nuclei by their electron clouds set by nature.

HADRONIC LAW 8.7.4-II: Hadronic fusion reactors should have means for the systematic and controlled coupling of nuclear spins either in planar singlet coupling (figure 1) or in axial triplet coupling (Figure 4). In fact, under other types of couplings, nuclei strongly repel each other.

HADRONIC LAW 8.7.4-III: The most probable nuclear fusions occur between light, natural and stable elements when used as hadronic fuels (see Sect. 8.4.5 for details).

HADRONIC LAW 8.7.4-IV: The energy used by hadronic reactors should be the minimal possible for the operation of all engineering components. In fact, any available energy exceeding the indicated threshold causes instabilities, resulting in a decrease in efficiency.

HADRONIC LAW 8.7.4-V: In addition to the verification of the above laws, hadronic fusion reactors should have a "trigger" forcing two nuclei to be in contact with each other so as to activate strong nuclear forces. The insufficiencies of Hadronic Laws I to IV to achieve controlled and sustained nuclear fusions were identified in the 2007 paper [8-80]. The unknown final mechanism to achieve nuclear fusions was then called the "trigger" (see Sect. 8.8 for its identification).

8.7.5. The new chemical species of magnecules

Evidently, Hadronic Law I on the systematic exposure of nuclei out of their electron clouds is a necessary pre-requisite for systematic nuclear fusions. Following a decade of mathematical, theoretical and experimental research under considerable private funds, Santilli presented in the 1998 memoir [8-62] new chemical species permitting indeed the needed nuclear exposure nowadays known as *Santilli magnecules* in order to distinguish them from the conventional chemical species of "molecules" (see books [4-9] [8-63] for a systematic treatment, samples of independent review [8-64] [8-65] and paper [8-66] for recent experimental verification).

By noting that the contaminants of fossil fuel exhaust (such as carbon monoxide, CO, and hydrocarbons, HC) are combustible, our environmental problems are primarily due to the strength of the valence bonds, which is such that they prevent their full combustion.

The solution proposed in Ref. [8-62] is the creation of the species of magnecules via a new bond between individual atoms which bond: (1) is stable under ambient conditions; (2) sufficiently weaker than the conventional valence bond (as a condition to achieve full combustion); and (3) is null at the combustion temperature (so as to increase the combustion temperature due to the absence of molecular separations, which are necessary for the combustion of molecules).

The new magnecular bond between atoms is based on the use of a magnetic field sufficiently strong to deform atomic electron clouds into toroids, which then bond to each other with opposite magnetic polarities North-South-North-South...illustrated in Figure 8-13. Hence, Santilli's magnecular bond is based on the creation within atomic structures of a magnetic moment that does not exist in the conventional spherical distribution of the same orbitals. Evidently, toroidal polarizations are, individually, unstable because the spherical distribution is recovered in nanoseconds following the removal of the external magnetic field.

Nevertheless, when two toroidal polarizations are bonded together by opposing magnetic polarities, spherical distributions are again recovered in nanoseconds following the removal of the external magnetic field, but for the bounded pair as a whole, resulting in this way the:

DEFINITION 8.7.5-I [8-62]: Santilli's magnecules are stable clusters consisting of individual atoms (H, C, O, etc.), dimers (OH, CH, etc.) and ordinary molecules (CO, H_2O, etc.) bonded together by opposing magnetic polarities originating from toroidal polarizations of the orbitals of atomic electrons and other polarization effects.

As established with extensive industrial use of the new fuel with magnecular structure, called *MagneGas* (chemical symbol MG), the magnecular bond verifies all three conditions (1) (2) (3) because stable, much weaker than the valence bond, and null at its Curie temperature, which is the combustion temperature, thus confirming its primary magnetic character.

Hadronic refineries used in the industry consist of a metal vessel capable of withstanding pressures up to 1000 psi; the vessel is filled up with the selected liquid feedstock and houses a DC electric arc among one or more pairs of electrodes that are submerged within the selected fluid and powered by an external AC-DC converter with at least 100 Kw.

The equipment includes means for the continuous recirculation of the liquid feedstock through the arc so that magnecular structures are continuously removed from the arc following their creation. See U. S. patents [8-68] [8-69] [8-70].

The equipment is equipped with remote automatic control mechanisms for all operations, including the DC arc, pressure, feedstock flow, feedstock temperature, and compression of the produced fuel, among others.

The most important fuels with magnecular structure [8-63] [8-71] are the following:

8.7.5-I. Magnegas (MG). This gaseous fuel is synthesized from petroleum-based liquids by adding water. Its most salient feature is the achievement of full combustion with no detectable CO or HC in the exhaust due to its flame temperature, which is about three times that of natural gas [8-72]. Additionally, magnegas exhaust contains the desired percentage of oxygen since it is set by the percentage of water added to the petroleum base feedstock (see the 14.6% O_2 reported in Fig. 18, p. 85 of [8-63]).

8.7.5-II. MagneHydrogen (MH). MH is a new form of hydrogen obtained from MG via Pressure Swing Adsorption (PSA) separation via the use of zeolites for the Helium due to the size of magnecular clusters. Its most important feature is that of being ordinary atomic Hydrogen when inspected with InfraRed Detectors (IRD), while its specific weight when actually measured, results in a multiple of that of Hydrogen. Additionally, the polarized character of MH clusters allows them to bond to container walls, preventing seepage. In view of these and other features, MH is an ideal replacement for conventional H2 in the aerospace and automotive industries [8-63] [8-73]-[8-76].

Figure 8-17. A view of the certification by Princeton Gamma Tech Instruments (PGTI) of the excess clean energy produced by ICNF hadronic reactors, showing on the left the PGTI team and on the right the certification of the lack of neutron emission via the detector Sam 934 (Sect. 8.7.7).

8.7.5-III. HHO. The name HHO was introduced by Santilli in the 2006 paper [8-77] to denote a magnecular, stoicheiometric and gaseous form of water, thus having the atomic composition of 2/3 Hydrogen and 1/3 Oxygen. Its main features are combustion without any need for Oxygen, instantly melting bricks at contact and other anomalous features, that is, features that cannot be quantitatively explained via quantum chemistry. This confirms the need for a magnecular structure for its very existence. Santilli HHO gas has been extensively studied as an additive to petroleum in internal combustion engines for improving exhaust emissions, with over 1000 citations of papers [8-77] to date.

8.7.6. Intermediate Controlled Nuclear Fusions (ICNF)

In our view, Low Energy Nuclear Syntheses (LENS) [8-78], also called "cold fusion," do indeed achieve nuclear fusions, but they have not reached industrial uses because of insufficient energy to power all engineering means for controllable and sustainable pro- duction of nuclear energy.

By contrast, High Energy Nuclear Fusions [8-79], also called “hot fusions,” can indeed be attained in the laboratory, although hot fusions as well have not achieved results of industrial significance, and none is in sight at this writing, due to uncontrollable instabilities at the initiation of the fusion process.

In view of the indicated insufficiencies at low and high energies, Santilli proposed in the 2008 paper [8-80] a new type of nuclear fusions under the name of *Intermediate Controlled Nuclear Fusions* (ICNF), namely, nuclear fusions with energy intermediate between those of the cold and hot fusions. ICNF were then studied and developed at the theoretical, experimental, and industrial levels in works [8-81]-[8-104] at the U.S. publicly traded company *Thunder Energies Corporation,* now the private Hadronic Energies Corporation. The main principle for the engineering realization of ICNF, as suggested by a decade of tests, is the use of a DC arc between carbon electrodes submerged within a gaseous feedstock. Therefore, the hadronic fuels are given by *Carbon* C(6, 12, 0) [8-105] and the selected gas.

8.7.6-A. The engineering of hadronic reactors

Hadronic reactors for the engineering realization of ICNF consist of a metal vessel containing a gaseous hadronic fuel at pressure traversed by internal carbon electrodes with remote means for the monitoring and control of the arc power, arc gap, gas pressure, gas temperature, vessel temperatures, gas flow through the arc; a heat exchanger; a variety of neutrons and other detectors; remote, monitoring and control panels of the various functions with automatic disconnect of all systems in the event of any deviation of the data from pre-set values [8-68]-[8-70].

In regards to manufacturing data, tests [8-80]-[8-85] were done via hadronic reactors comprising cylindrical metal vessels with outside diameters ranging from 1 foot to 2 feet and lengths ranging from 2 feet to 6 feet, said vessels being certified to withstand internal pressures at least up to three times the expected operating pressure; electrodes fabricated form cylindrical graphite rods ranging from 1 inch to 2 inches diameter with the non-consuming anode generally being 4 inches long and the consuming cathode generally being a minimum 6 inches long; a metal jacket surrounding said vessel containing a coolant (such as water) which is recirculated through a heat exchanger; specially designed control panels for the various functions; and other engineering means.

8.7.6-B. Controllability of hadronic reactors.

Rather important for the appraisal of the industrial value of ICNF hadronic reactors is the understanding of the controlled character of nuclear fusion, which is achieved via the control of AC-DC power, electrode gap, hadronic fuel pressure, and other means.

8.7.6-C. Sustainability of hadronic reactors

Equally important is the understanding of the sustainability of the ICNF process. The only consumable parts of hadronic reactors are the carbon electrodes, which must be replaced when they are consumed. However, Santilli U. S. patents [8-58] [8-69] [8-70] provide various solutions via electrode pairs to assure continuous operations of hadronic reactors without time lost for electrode replacement.

8.7.6-D. Safety of hadronic reactors

It is also important to know the safety of hadronic reactors for operators and bystanders, which is provided by the documented lack of any emission of harmful radiation or release of radioactive waste; the automatic disconnect of the power for any surge of energy, temperature or pressure; the very nature of Santilli's nuclear fusions which can only occur in parts per million per mole of hadronic gaseous fuels [8-33], by therefore preventing explosions.

8.7.6-E. Protocol tests

All tests [8-80]-[8-85] were conducted using a 50 kW Miller Electric Dimension 1000 AC-DC converter operating at 40 kW, thereby supplying 0.866 kW per minute. The reader should be aware that the use (in lieu of a commercially available AC-DC converter) of the special DC power unit for the neutron synthesis provides a significant increase in the energy output due to special features of the DC arc not outlined in Sect. 8.5.

Samples of the solid and gaseous hadronic fuels were taken for all ICNF tests outlined under the Trail of Custody by the technician Jim Alban before and after the tests and sent out for analysis by independent companies.

More specifically, laboratory bottles were filled with the gaseous hadronic fuel at an ambient temperature of approximately 80°F (26°C) before and after the reactor was activated. Said samples were individually marked and shipped to *Oneida Research Services* (ORS) of Whitebnodo, New York, for the *two* different analyses of reports [8-94]-[8-100], the first for *molecular* counts and the second for *nuclear* counts with Atomic Mass Units from 2 u to 400 u, which were done via an Internal Vapor Analyzer, model 110-s operated with SOP MEL-1070. The generally low pressure of the various laboratory bottles was also marked and is reported in the individual ORS reports.

The following information is important for the proper interpretation of ORS reports: 1) The molecular and nuclear counts are inequivalent due to magnecular anomalies caused by the DC arc (Sect. 8.7.5). 2) The total number of nuclear counts before and after the tests. Also under the Trail of Custody by Jim Alban, samples of the graphite used for the electrodes were taken before and after each test, marked and shipped to *Constellation Technology* of Largo, Florida, for analyses available in reports [8-101]-[8-104]. It should be noted that the latter analyses are for *solid* traces of new elements in the electrodes, following tests that confirm some of the nuclear fusions detected by the ORS analyses but not all since primary nuclear fusions occur at the gaseous level.

8.7.6-I. ICNF with the Nitrogen hadronic reactor [8-81].

The Nitrogen hadronic reactor [8-81]-[8-83] and Figures 8-14, 8-15 were built according to the specifications described in Section 9.7.6A. In particular, the metal vessel was constructed from a Schedule 40 steel tube, measuring 1 foot × 2 feet, with a 1/2-inch thickness, weighing 325 pounds, plus side flanges weighing 125 pounds each, for a total of 575 pounds, certified to withstand 300 pounds per square inch (psi). All tests were done with gaseous hadronic fuels at 100 psi and for a maximum of *two minutes* due to the rapidity of the temperature surge.

Among the various tests conducted with the Nitrogen hadronic reactor from 2005 to 2016, we outline the following tests below, with the understanding that to avoid excessive length, all technical details are referred to [8-81]-[8-83].

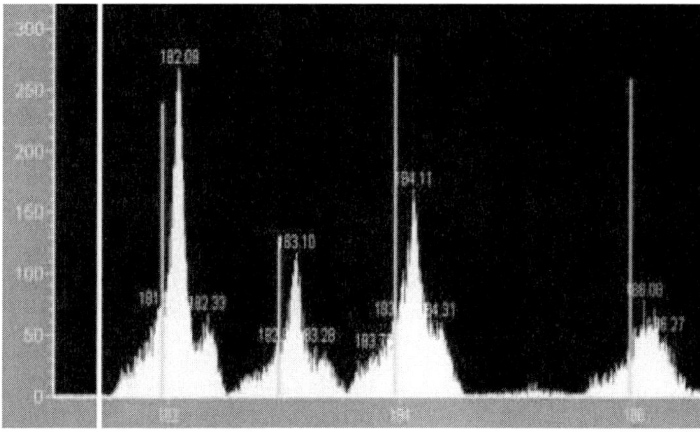

Figure 8-18. A view of the experimental verification of Santilli pseudo-nuclei via a series of modified Tungsten isotopes done at the Ural Federal University, Yekaterinburg, Russia (Sect. 8.8.4 and Ref. [8-106)

8-7-6-II. ICNF with Deuterium and Carbon.

The Nitrogen hadronic reactor was filled with commercial-grade deuterium gas at a pressure of 100 psi and under a 40 kW DC power supply. Following two minutes of operation, the external temperature of the reactor went from 26° C to 150° C. Two laboratory bottles, before and after the activation of the reactor, were filled with the gas at the pressure indicated in the reports, labeled HCN1 and HCN2, respectively, and shipped to ORS. The results are available in Ref. [8-94] and will be analyzed for nuclear fusions in Sect. 8-8.

8-7-6-III. ICNF with Hydrogen and Carbon.

Tests confirmed the preceding results in the Nitrogen hadronic reactor with a commercial grade Hydrogen at 100 psi pressure and 40 kw DC power, as reported in Sect. 6 of [8-81] and in ORS report [8-96] for bottles market HC1 and HC2. An important result of this test is that under the same conditions of pressure, power, electrodes, etc., of the preceding test with Deuterium and Carbon, the operation with hydrogen gas produced an energy excess bigger than that with Deuterium gas since in two minutes of operation the temperature of the exterior wall of the reactor went from 26^oC to 254° C with about 1.72% *increase* of the temperature compared to the test with Deuterium.

8-7-6-IV. ICNF with Magnegas and Carbon.

The most successful tests with the Nitrogen hadronic reactor occurred when magnegas, a gaseous fuel with a novel magnetic structure [5-9], was used as a hadronic fuel [8-70]. The results of the ORS analyses are reported in [8-96] for bottles marked MG1 and MG2. A main result of various tests is that the Nitrogen hadronic reactor operating with magnegas at 100 psi pressure and 40 kW power increased from 26 °C to 254 °C in one minute, rather than the two minutes required for the Hydrogen-Carbon tests, thus implying a 3.44-fold increase in efficiency compared to the Deuteron-Carbon tests.

We should indicate the conduct of additional tests with the Nitrogen hadronic reactor using various gaseous fuels, whose analyses are available from [8-95].

8.7.6-V. ICNF with the Oxygen hadronic reactor

The Oxygen hadronic reactor (see [8-82]-[8-83], independent studies in [8-89], ORS reports [8-97]-[8-98] and Fig. 8-15) was built in 2010 for testing the ICNF of Helium and Carbon into Oxygen, by therefore using Helium as the gaseous hadronic fuel.

The hadronic reactor comprised a vertical 1 foot × 4 feet Schedule 40 steel cylinder certified to withstand 500 psi; a chamber surrounding said vessel for flowing water as a coolant; the flow of the gaseous hadronic fuel through the electrodes for its cooling; and the remaining engineering component illustrated in Fig. 5.

The reactor was additionally built to test the feasibility of the new principle of combustion, subsequently released in 2018 under the name Hyper Combustion [8-107], which is intended to achieve the full combustion of fossil fuels via a combination of conventional combustion and ICNF in Parts Per Million by volume (ppmv).

For the test conducted in April 2010, the laboratory bottles, both before and after the test, were filled with the internal gas at the pressure indicated in the reports, market HT1 and HT2, and shipped to ORS for analysis. The results of this analysis are available from [8-97]. The tests were repeated in February 2011, and market HE1 and HE2 were sent to ORS for analysis, whose results are available from [8-98] (see Sect. 8-8 for their analysis).

The primary result of the tests was the proof, under various eyewitnesses, that when filled up with Helium at 100 psi and operated with a 40 kw AC-DC converter, the Oxygen hadronic reactor did indeed produce a steam-sustainable for two minutes, after which the cooling system was insufficient to maintain the reactor at a constant temperature.

During additional tests done on May 15, 2011, with the Helium hadronic fuel at 150 psi, in two minutes of operation, the mixture of Helium and synthesized Oxygen sent the internal temperature gauge off the 10000° C limit and melted the top Helium recirculation port, with an impressive release of the incandescent interior gas after which all tests with the Oxygen hadronic reactor were terminated for safety. The technicians (who eye witnessed the discharge at a distance) nicknamed the hadronic reactors "Dragons" (Dragons I, II, and III for the Nitrogen, Oxygen, and Silicon reactors, respectively).

8.7.6-VI. ICNF with the Silicon hadronic reactor

The Silicon hadronic reactor (see [8-82] [8-83], video [8-85] and Fig. 816) was built to test the ICNF of Oxygen and Carbon into Silicon via the use of *air* as hadronic fuel instead of pure Oxygen because the natural mixture of 78% Nitrogen and 21% Oxygen is known to quench the Oxygen reactance experienced in Section 4.5.

The hadronic reactor consisted of a Schedule 40 steel tube with a 1-foot diameter and 6 feet in length, certified to withstand a pressure of 5,000 psi. Air was continuously pumped through the reactor at 1000 psi. A 50 kw AC-DC converter powered the arc. The reactor was surrounded by a jacket in which water was continuously pumped at ambient pressure. The superheated air and cooling water from the reactor were mixed to power an electricity-producing turbine, whose data are analyzed in Sect. 8.8. To avoid excessive length, we suggest that interested readers view the video [8-85] for a detailed description of this third hadronic reactor, including the identification of the various members of the experimental team

The analyses for the gaseous part of the test are available from ORS [8-99] [8-100], and the analyses for the solid part are available from Constellation Technology [8-101]-[8-104]. The main result of the tests with the Silicon hadronic reactor is that under various eyewitnesses (see Fig. 8-16 and Ref. [8-85]), *the Silicon hadronic reactor did prove the capability of ICNF to produce clean excess energy for* 15 m *(fifteen minutes),* after which the remote monitoring and control panels automatically disconnected the operation for the inability of the cooling system to maintain the reactor temperature within a pre-set safety value.

No additional tests were done with the Silicon hadronic reactor due to a lack of funds for the construction of a properly engineered *prototype hadronic power plant.*

8.7.7. Independent certifications of clean excess energy produced by ICNF hadronic reactors

8.7.7-I. Certification by Princeton Gamma Tech Instruments

On February 12, 2011, L. Ying, President of Princeton Gamma-Tech Instruments (PGTI), Princeton, N.J., and his associates R. Brenna and T. R. Kuliczkowski, visited the laboratory of Thunder Energies Corporation in Tarpon Springs, FL, to verify the excess of clean energy produced by Santilli Nitrogen Reactor (Figure 8-14) for the synthesis of Nitrogen- 14 from Carbon-12 and Deuterium-2 under the operating conditions indicated in Sect. 8.7.6-I. The results of the tests were reported in papers [8-86]-[8-89].

Thanks to the use of their neutron and other detectors (Figure 8-17), PGTI experimentalists certified that the Santilli hadronic reactor 8.7.6-I emits no neutrons or other massive particles. Subsequently, they filled up a laboratory bottle of the internal gas and certified that the hadronic reactor released no radioactive waste.

To measure the produced excess energy, temperature-resistant sensors were securely fastened to the surfaces of the reactor and one of the endplates. Temperature readings were noted down each minute of operation to produce a thermal profile of the reactor. A thermal Finite Element Analysis (FEA) was conducted to estimate the temperature rise. The comparison of the measured thermal profile against the FEA computed values at 5 MJ indicated the generation of excess heat above the total used energy input of 5.4 MJ, which, in two minutes of operation, corresponded to the production of a micro-mole of synthesized Nitrogen with 10 MeV of clean thermal energy per each Nitrogen synthesis.

Additionally, a mass spectroscopic analysis was performed on gas samples extracted before and after ignition. Their comparison established the magnecular character of the treated gas (Sect. 8.7.5).

8.7.7-II. Certification by Y. Yang et al.

On January 18, 2013, the Canadian experimentalist Y. Yang and his associates, J.V. Kadeisvili and C. Lynch, visited the laboratory of Thunder Energies Corporation in Tarpon Springs, FL, and conducted a series of tests with Santilli's ICNF hadronic reactor 8.7.6-I for the synthesis of Nitrogen-14 from Carbon-12 and Deuterium-2 and published their results in a paper [8-90].

To validate the performance, Santilli's ICNF hadronic reactor was powered by a Miller Dimension 1000 AC-DC converter, delivering approximately 36 kW (40 kVA) of power to the carbon electrodes at settings of 40 VDC and 900 A. The reactor chamber was filled with 99.999% pure deuterium gas, creating an optimal environment for fusion reactions.

Thermal analyses conducted during the experiments revealed the generation of excess heat, at least ten times the 36 kW input, within two minutes of operation. This excess energy production confirms Santilli's theoretical predictions and experimentations of the ICNF fusion of Carbon-12 and Deuterium-2 into Nitrogen-14.

Measurements during the tests confirmed the absence of harmful radiation emissions or radioactive waste release, validating the safety of the ICNF process.

The independent certification by Yang and his associates, Kadeisvili and Lynch, substantiates the claims regarding Santilli's ICNF hadronic reactors. Their rigorous experimentation underscores the potential of ICNF technology as a clean and efficient energy source, paving the way for further research and potential industrial applications.

8.7.7-III. Additional certification of clean excess energy

It should be indicated that the thermal data, chemical analysis and spectroscopic measurements of certifications [8-86]-[8-90] have been analyzed in detail and found to be correct by a number of independent scholars, such as R. B. Lanjewar [8-91], I. B. Das Sarma [8-92], U. Abundo [8-93], and others.

8.8. Santilli hyperFusions (HF)

8.8.1. The birth of HyperFusions

The ICNF tests were conducted from 2015 to 2016 via the Nitrogen, Oxygen, and Silicon hadronic reactors, as reported in Sect. 8.7.6, were conceived under the assumption that the implementation of Hadronic Laws I-V for nuclear fusion required the yet unknown realization of a trigger. It was then surprising for the author to see the production of clean nuclear energy as a significant multiple of the used energy.

Subsequent studies revealed that, despite the systematic exposure of nuclei out of their electron clouds, despite the correct spin and magnetic couplings and other features (Figure 13), the sole creation of a magnecular structure in gaseous hadronic fuels is insufficient for a representation of the amount of energy output of the ICNF nuclear fusions reported by the ORS Laboratory due to the extremely strong Coulomb repulsion between natural nuclei.

It was then rather natural to assume for the first time in the 2022 paper [8-33] that the nuclear fusions reported by the ORS Laboratory were due to the prior synthesis by a sufficiently powerful DC arc of the pseudo-nuclei of Sect. 8.3 since natural nuclei attract them with a very strong Coulomb force under which the physical contact between natural nuclei and pseudo-nuclei is unavoidable with ensuing inevitable fusion illustrated in Figure 8-7. Equivalently, we can say that the industrial realization of the "trigger" of Hadronic Law 8.7.4-V is given by the controlled and sustainable synthesis of pseudo-nuclei by the plasma of a DC arc.

Let us recall again from Sections 1.1, 8.6 and 8.7 that, to prevent insidious inconsistencies that generally remain undetected by quantum mechanical experts (Chapter 5), the consistent study of manifestly irreversible HyperFusions requires the use of the Lie-admissible branch of hadronic mechanics with basic time evolution given by the Heisenberg-Santilli genoequation $idA/dt = (A, H) = A < H — H > A = ARS — HSA$, where R and $S \neq R$ are the forward and backward Santillians, in the standard realization with $R = 1$ and $S = 1 — F/H$ given by $idA/dt = (A, H) = [A, H]^* + FA$, where: H is the Hamiltonian representing potential interactions, S is the forward Santillian representing the extended character of nucleons and their consequential contact non-potential interactions; and F represents the historical Lagrange's and Hamilton's external terms originating irreversibility (Chapter 3), with related mathematics (Chapter 4), hadronic uncertainties (Chapter 6) and relativities (Chapter 7).

8.8.2. Insufficiency of calorimetric measurements.

As indicated in the preceding section, the best excess energy produced by the Nitrogen hadronic reactor over the used energy in two minutes of operation, which was appraised via calorimetric measurements to be $\Delta E = E^{out} - E^{in} = 23.656 - 0.666$ kwh = 2.99 kwh [8- 81] [8-86] [8-89], did establish the controlled and systematic production of excess energy by ICNF hadronic reactors.

Nevertheless, these appraisals were significantly *below* the excess energy actually produced because they were based on calorimetric measurements of the external (rather than the actual internal) temperature of the hadronic reactors.

As an example, the use in the hadronic reactor of Section 8.7.6-I of the Helium as hadronic fuel at 150 psi did indeed yield a multiple of the preceding appraisal but was still insufficient for full calorimetric internal measurements of excess heat and energy.

8.8.3. Experimental confirmation of pseudo-nuclei by the Ural Federal University, Yekaterinburg, Russia

Just about the time of completing the writing of this chapter (February 24, 2025), the author was informed that the experimentalists M. Kashchenko, M. Kovalenko, V. Pechorsry and N. Kashchenko of the Ural Federal University, Yekaterinburg, Russia, have achieved the first known experimental confirmation of the existence of pseudo-nuclei ([8-33] and Sect. 8.3) via the detection of a number of pseudo-Tungsten isotopes created by a DC electric arc using Tungsten for one of the two electrodes (Figure 9-19 and Ref. [8-106]).

8.8.4. Clean excess heat from detected nuclear fusions

In view of the indicated insufficiencies of thermal calculations of excess heat produced by hadronic reactors, in this section, we report the excess energy output presented in paper [8-33] as direct calculations of the energy produced by the following representative count per part per million per volume (ppmv) of nuclear fusions with atomic mass de- noted with the letter u (ORS laboratories [8-94]) (a)Δ 2 u: 18, 550, 801 - 16, 075, 402 = 2, 475, 399 ppmv, (b)Δ 3 u: 41, 165 — 30, 269 = 10, 896 ppmv, (c)Δ 4 u = 76 - 0 = 76 ppmv, (d)Δ 14 u : 3, 555 - 2, 841 = 714 ppmv, (e)Δ 16 u : 3, 010 - 1, 205 = 1, 805 ppmv, (f)Δ 18 u 2, 949 - 2, 718 = 231 ppmv, (g)Δ 28 u: 30, 171 - 24, 684 = 3, 687 ppmv.

By identifying the actual nuclear fusion per count, the above ORS data imply the following corresponding energy production (see Ref. [9-33] for details and the Table of Nuclides [8-105]) $E_a = 2.014102$ u, $E_b = 4.002603$ u, $E_c = 12.00000$ u, $E_d = 14.003074$ u, $E_e = 15.994915$ u, $E_f = 15.004915$ u, $E(g) = 27.976927$u for an approximate total in parts per millions per mole of 54, 546, 518 MeV $= 2.43 \times 10^{-12}$ kwh.

By assuming that the gas in the reactor is a perfect gas, by assuming the related law $PV = nRT$ and by recalling that one mole contains 6.02×10^{23} particles, that is, 6.02×10^{17} *millions* of particles, the total energy output calculated in paper [8-33] is given by $\Delta E = E^{out} — E^{in} = 2.43 \times 10—12 \times n \times 6.02 \times 10^{17} — 1.333$ kwh $\approx n \times 14.628 \times 10^5$kwh where the number of moles of hadronic fuel in the interior of the hadronic reactor is given by (see paper [8-33] for details) $n \approx 7 \times 10^{-5}$.

Consequently, the approximate total output of controlled and sustained clean energy of the considered ICNF is given by $\Delta E = E^{out} - E^{in} = 7 \times 10^{-5} \times 15 \times 10^5 - 1.333$ kwh ≈ 100 kwh under the operating conditions set in Section 8.7.6-VI (50 kwh for AC-DC converter, gaseous hadronic fuel set at 1000 psi and operations for 15 m), out of which 70 kwh are predicted to be lost in the conversion of heat to electricity, by remaining with 30 kwh of clean electricity produced by a modular reactor sufficient for households or small factories.

It should be finally indicated that, under the engineering conditions of Sect. 8.7.6, HyperFusions cannot produce atomic explosions because, in view of inevitable random processes in any plasma, they are naturally restricted to occur in parts per million per volume of one mole (ppmv).

References of Chapter 8

[1] Rutherford, E.: The Existence of a Neutron, Bakerian Lecture: Nuclear Constitution of Atoms. *Proc. Roy. Soc. A* 97, 374-382 (1920), royalsocietypublishing.org/doi/10.1098/rspa.1920.0040

[2] Chadwick, J.: *Proc. Roy. Soc. A*, 136, 692-723 (1932), royalsocietypublishing.org/doi/10.1098/rspa.1932.0112

[3] Pauli, W.: Rapports du Septième Conseil de Physique Solvay, *Gauthier-Villars,* Paris, 324 (1933), www.worldcat.org/title/23422639?oclcNum=23422639

[4] Fermi, E.: *Nuclear Physics,* University of Chicago Press (1949).

[5] Wietfeldt, F. E.: Measurements of the Neutron Lifetime, *Atoms* 6, 1-19 (2018), https://www.mdpi.com/2218-2004/6/4/70

[6] Norman, R. and Dunning-Davies, J.: Hadronic paradigm assessed: neutroid and neutron synthesis from an arc of current in hydrogen gas, *Hadronic Journal* 40, 119-132 (2017), https://www.santilli-foundation.org/docs/norman-dunning-davies-hj.pdf

[7] Borghi, C., Giori, C. and Dall'Olio, A.: Communications of CENUFPE, Number 8 (1969) and 25 (1971), reprinted in *Phys. Atomic Nuclei* 56, 205-221 (1993).

[8] Santilli, R. M.: Apparent consistency of Rutherford's hypothesis on the neutron as a compressed hydrogen atom, *Hadronic J.* 13, 513-533 (1990), http://www.santilli-foundation.org/docs/Santilli-21.pdf

[9] Santilli, R. M.: Apparent consistency of Rutherford's hypothesis on the neutron structure via the hadronic generalization of quantum mechanics, nonrelativistic treatment. ICTP communication IC/91/47, (1992), https://www.santilli-foundation.org/docs/Santilli-150.pdf

[10] Santilli, R. M.: The synthesis of the neutron according to hadronic mechanics and chemistry. *Journal Applied Sciences* 5, 32-47 (2006).

[11] Santilli, R. M.: Recent theoretical and experimental evidence on the synthesis of the neutron, Communication of the JINR, Dubna, Russia, No. E4-93-252 (1993).

[12] Santilli, R. M.: Recent theoretical and experimental evidence on the synthesis of the neutron, *Chinese J. System Eng. and Electr.* 6, 177-195 (1995), https://www.santilli-foundation.org/docs/Santilli-18.pdf

[13] Santilli, R. M.: *The Physics of New Clean Energies and Fuels According to Hadronic Mechanics,* Special issue, *Journal of New Energy* (1998), https://www.santilli-foundation.org/docs/Santilli-114.pdf

[14] Santilli, R. M.: Reduction of Matter in the Universe to Protons and Electrons via the Lie-isotopic Branch of Hadronic Mechanic. *Progress in Physics,* 19, 73-99 (2023), https://www.ptep-online.com/2023/PP-65-09.PDF

[15] Santilli, R. M.: Apparent confirmation of Don Borghi's experiment on the laboratory synthesis of neutrons from protons and electrons, *Hadronic J.* 30, 29-41 (2007),

https://www.santilli-foundation.org/docs/Santilli-114.pdfwww.i-b-r.org/NeutronSynthesis.pdf

[16] Santilli, R. M.: Confirmation of Don Borghi's experiment on the synthesis of neutrons. *arXiv* physics/0608229v1 Aug.15, 2007,

[17] arxiv.org/pdf/physics/0608229v1.pdf

[18] Burande, C. S.: On the experimental verification of Rutherford-Santilli neutron model, *AIP Conf. Proc.* 158, 693-721 (2013), https://www.santilli-foundation.org/docs/Burande-2.pdf

[19] Santilli, R. M. and Nas, A.: Confirmation of the Laboratory Synthesis of Neutrons from a Hydrogen Gas. *Journal of Computational Methods in Sciences and Eng.* 14, 405–414 (2014), https://www.hadronictechnologies.com/docs/neutron-synthesis-2014.pdf

[20] Santilli, R. M.: Apparent Nuclear Transmutations without Neutron Emission Triggered by Pseudoprotons. *American Journal of Modern Physics*, 4, 15-18 (2015), https://www.sciencepublishinggroup.com/journal/paperinfo.aspx?journalid=122&doi=10.11648/-j.ajmp.20150401.13

[21] Haan, V. de.: Possibilities for the Detection of Santilli Neutroids and Pseudo- protons, *American Journal of Modern Physics* 5, 131-136 (2015), https://www.santilli-foundation.org/docs/Pseudo-proton-deHaan.pdf

[22] Norman, R., Beghella Bartoli, S., Buckley, B., Dunning-Davies, J., Rak, J. and Santilli, R. M.: Experimental Confirmation of the Synthesis of Neutrons and Neutroids from a Hydrogen Gas, *American Journal of Modern Physics* 6, 85-104 (2017), https://www.santilli-foundation.org/docs/confirmation-neutron-synthesis-2017.pdf

[23] Driscoll Bohrs, R. B.: Atom Completed: the Rutherford-Santilli Neutron, *APS Conf. Proc.* APR 03, April 5-8, (2003), ui.adsabs.harvard.edu/abs/2003APS.APR.D1009D/abstract

[24] [8-23] Chandrakant, S. and Burande, C. S.: On the Rutherford-Santilli Neutron Model. *AIP Conf. Proc.* 1648, 51000-1 - 51000-6 (2015), https://www.santilli-foundation.org/docs/1.4912711(CS-Burande(1)).pdf

[25] Kadeisvili, J. V.: The Rutherford-Santilli Neutron. *Hadronic J.* 31, 1-125 (2008), https://www.i-b-r.org/Rutherford-Santilli-II.pdf Website http://www.i-b-r.org/Rutherford-Santilli-neutron.htm

[26] Burande, C. S.: Santilli Synthesis of the Neutron According to Hadronic Mechanics. *American Journal of Modern Physics* 5, 17-36 (2016), https://www.santilli-foundation.org/docs/pdf3.pdf

[27] Beghella-Bartoli, S.: Significance for the EPR Argument of the Neutron Synthe- sis from Hydrogen and of a New Controlled Nuclear Fusion without Coulomb Barrier. Proceedings of the *2020 Teleconference on the EPR argument, Curran Associates Conference Proceedings,* New York, USA, 459-466 (2021).

[28] Dunning-Davies, J.: A Present Day Perspective on Einstein-Podolsky-Rosen and its Consequences, *Journal of Modern Physics* 12, 887-936 (2021), https://www.scirp.org/journal/paperinformation.aspx?paperid=109219

[29] Santilli, R. M.: Studies on A. Einstein, B. Podolsky and N. Rosen prediction that quantum mechanics is not a complete theory, III: Illustrative examples and applications, *Ratio Mathematica* 38, 139-222 (2020), https://eprdebates.org/docs/epr-review-iii.pdf

[30] Santilli, R. M.: The etherino and/or the neutrino Hypothesis? *Found. Phys.* 37, 670-695 (2007), http://www.santilli-foundation.org/docs/EtherinoFoundPhys.pdf

[31] Santilli, R. M.: Nuclear Realization of Hadronic Mechanics and the exact representation of nuclear magnetic moments, *Communications in Mathematical and Theoretical Physics* 2, 1-66 (1999), https://santilli-foundation.org/docs/santilli-519.pdf

[32] Santilli, R. M.: Lie-isotopic representation of stable nuclei II: Exact and time-invariant representation of the Deuteron data, *Ratio Mathematica* 52, 64-130 (2024) ttps://www.santilli-foundation.org/docs/4-RM-SI-Vol-52-2024-fin.pdf

[33] Santilli, R. M.: Lie-isotopic representation of stable nuclei III: Exact and time-invariant representation of nuclear stability, *Ratio Mathematica* 52, 131-175 (2024), https://www.santilli-foundation.org/docs/5-RM-SI-Vol-52-2024-fin.pdf

[34] Santilli, R. M.: Apparent Resolution of the Coulomb Barrier for Nuclear Fusions Via the Irreversible Lie-admissible Branch of Hadronic Mechanics, Progress in Physics, 18, 138-163 (2022), http://www.santilli-foundation.org/hyperfusion-2022.pdf

[35] Burande, S: Santilli Synthesis of the Neutron According to Hadronic Mechanics, *American Journal of Modern Physics* 5(2), 37-41 (2016), https://www.santilli-foundation.org/docs/HMMC.pdf

[36] Dhondge, S. S.: Studies on Santilli Three-Body Model of the Deuteron According to Hadronic Mechanics, *American Journal of Modern Physics* 5(2), 46-52 (2016), https://www.santilli-foundation.org/docs/HMMC.pdf

[37] Bhalekar, A. A. and Santilli, R. M.: Exact and Invariant representation of nuclear magnetic moments and spins according to hadronic mechanics, *American Journal of Modern Physics* 5, 56 (2016), https://www.santilli-foundation.org/docs/pdf5.pdf

[38] Santilli, R. M.: A quantitative isotopic representation of the Deuteron magnetic moment, in *Proceedings of the International Symposium Dubna Deuteron-3e,* Joint Institute for Nuclear Research, Dubna, Russia (1994), http://www.santilli-foundation.org/docs/Santilli-134.pdf

[39] Santilli, R. M.: Iso-Representation of the Deuteron Spin and Magnetic Moment via Bohm's Hidden Variables, *Progress in Physics* 18, 74-81 (2022), https://www.santilli-foundation.org/docs/pip-3.pdf

[40] Santilli, R. M. and Sobczyk, G.: Representation of nuclear magnetic moments via a Clifford algebra formulation of Bohm's hidden variables, *Scientific Reports* 12, 1-10 (2022), http://www.santilli-foundation.org/Santilli-Sobczyk.pdf

[41] Dhondge, S.: Santilli's Hadronic Mechanics Formation for the Deuteron, *AIP Conference Proceedings* 648, 510009 (2015), http://www.santilli-foundation.org/docs/1.4912714(SS-Dhondge).pdf

[42] Muktibodh, A. S.: Studies on Santilli's lsonumber Theory *American Journal of Modern Physics* 5, 17-36 (2016), https://www.santilli-foundation.org/docs/pdf4.pdf

[43] Dhondge, S.: Studies on Santilli Three-Body Model of the Deuteron According to Hadronic Mechanics, *American Journal of Modern Physics* 5, 37(-452016), https://www.santilli-foundation.org/docs/deuteron-2018.pdf

[44] Thunder Energies Corporation, Critical Issues in Detecting Smuggled Fissionable Materi http://www.i-b-r.org/docs/Detection-fissionable-material.pdf

[45] Thunder Energies Corporation, Specifications of the Directional Neutron Source model DNS-3B-5 (2019), http://www.i-b-r.org/docs/DNS-brochure-9-12-19.pdf Short movie on the operation of the DNS http://www.i-b-r.org/docs/MagnaPower.mp4 neutron counts by the Ludlum detector 375 http://www.i-b-r.org/docs/Ludlum-Alarms.mp4 Confirmation by the Berke- ley Nucleonics neutron detector SAM 940 http://www.i-b-r.org/docs/Sam-Alarms.mp4

[46] Santilli, R. M.: Apparent Experimental Confirmation of Pseudoprotons and their Application to New Clean Nuclear Energies. *Intern. Journal of Appl. Phys. and Math.* 9, 72-100 (2019), https://www.santilli-foundation.org/docs/pseudoproton-verification-2018.pdf

[47] Santilli, R. M.: An intriguing legacy of Einstein, Fermi, Jordan and others: The possible invalidation of quark conjectures, *Found. Phys.* 11, 384-396 (1981), https://www.santilli-foundation.org/docs/Santilli-36.pdf
[48] Santilli, R. M.: Inconsistencies of Neutrino and Quark Conjectures and their Negative Environmental Implications Semantic Scholar March 2006, https://arxiv.org/abs/physics/0603262
[49] Hoyle, F.: A New Model for the Expanding Universe, Monthly *Notices of the Royal Astronomical Society* 108, 372-382 (1948), https://doi.org/10.1093/mnras/108.5.372
[50] Kikawa, T.: *Measurement of Neutrino Interactions and Three Flavor Neutrino Oscillations in the T2K Experiment,* Springer Nature (2016).
[51] American Chemical Society, Energy from the Sun, website, https://www.acs.org/content/acs/en/climatescience/energybalance/ener gyfromsun.html
[52] Oberauer, L., Lanni, A. and Serenelli, A.: *Solar Neutrino Physics,* Wiley (2020). [8-52] Kmane, A. S.: *Introductory Nuclear Physics,* Wiley (2008).
[53] Encyclopedia Britannica, Structure of the nucleus, https://www.britannica.com/science/atom/Structure-of-the-nucleus
[54] NEI, What Happens to Nuclear Waste in the U.S.?, website, https://www.nei.org/news/2019/what-happens-nuclear-waste-us
[55] Santilli, R. M.: Hadronic energy, *Hadronic J.* 17, 311-225 (1994), http://www.santilli-foundation.org/docs/hadronic-ener gy.pdf
[56] Santilli, R. M.: Conceivable new recycling of nuclear waste by nuclear power companies in their plants, *arxiv*, 9704014 (1997), https://arxiv.org/pdf/physics/9704014.pdf
[57] Dunning-Davies, J.: A Present Day Perspective on Einstein-Podolsky-Rosen and its Consequences, *Journal of Modern Physics* 12, 887-936 (2021),
[58] Santilli, R. M.: Problematic aspects of 20th-century antiparticles and their apparent resolution via isodual mathematics, *Ratio Mathematica* 52, 188-208 (2024), https://www.santilli-foundation.org/docs/7-RM-SI-Vol-52-2024-fin.pdf
[59] Tsagas, N. F., Mystakidis, A., Bakos, G., Sfetelis, L., Koukoulis D. and Trassani- dis, S.: Experimental verification of Santilli's clean subnuclear hadronic energy, *Hadronic Journal* 19, 87-90 (1996), http://www.santilli-foundation.org/docs/N-Tsagas-1996.pdf http://www.santilli-foundation.org/docs/Thessaloniki-Tsagas-experiment.ppt

[60] Santilli, R. M.: Apparent Nuclear Transmutations without Neutron Emission Triggered by Pseudoprotons. *American Journal of Modern Physics*, 4, 15-18 (2015), https://santilli-foundation.org/docs/santilli-521.pdf

[61] Santilli, R. M.: The Physics of New Clean Energies and Fuels According to Hadronic Mechanics, Special issue of the Journal of New Energy (1998), https://www.santilli-foundation.org/docs/Santilli-114.pdf

[62] Santilli, R. M.: Theoretical prediction and experimental verification of the new chemical species of magnecules, Hadronic Journal 21, 789-894 (1998), http://www.santilli-foundation.org/docs/Santilli-43.pdf

[63] Santilli, R. M.: *The New Fuels with Magnecular Structure,* International Academic Press (2008), http://www.i-b-r.org/docs/Fuels-Magnecular-Structure.pdf Italian translation by G. Bonfanti http://www.santilli-foundation.org/docs/Carb-Strutt-Magnecolare.pdf

[64] Tandge, V.: Advances in hadronic chemistry and its applications," Springer Nature Foundations of Chemistry Vol. 7, 163–179 (2015), https://doi.org/10.1007/s10698-015-9218-z

[65] Pandhurneka, C. P. and Zodape S. P.: Santilli's Magnecules and Their Applications, American Journal of Modern Physics); 6(4-1): 64-773 (2017), http://www.santilli-foundation.org/docs/magnecules-2017.pdf

[66] Yang, Y. Kadeisvili, J. V. and Marton, S.: Experimental Confirmations of the New Chemical Species of Santilli Magnecules, The Open Physical Chemistry Journal 5, 1 (2013), https://www.santilli-foundation.org/docs/Magnecules-2012.pdf

[67] Aringazin, A. K.: Toroidal configuration of the orbit of the electron of the hydrogen atom under strong external magnetic fields, Hadronic J. 24, 134 (2001), http://www.santilli-foundation.org/docs/landau.pdf

[68] Santilli, R. M.: Apparatus and method for recycling contaminated liquids. U. S. patent No. 6,540,966 pdfpiw.uspto.gov/.piw?docid=06540966

[69] Santilli, R. M.: Durable and efficient equipment for the production of combustible and non-pollutant gas from underwater arcs and method, therefore. U.S.patent No. 6,183,604 patentimages.storage.googleapis.com/4c/ad/2b/874b5ffe4a2f12/US6183604.pdf

[70] Santilli, R. M.: Method and Apparatus for the industrial production of new hydrogen-rich fuels. U. S. patent No. 9,700,870,B2 http://www.santilli-foundation.org/docs/Magnecule -patent.pdf

[71] Santilli, R. M.: The new fuels with magnecular structure and their environmental applications, lecture, http://www.world-lecture-series.org/santilli-new-fuels

[72] Alfano, R. R.: CCNY Certification of Magnegas Flame Temperature, http://www.santilli-foundation.org/docs/Magnegas

[73] Alfano, R. R.: CCNY Certification of Magnegas Flame Temperature, Report, http://www.santilli-foundation.org/docs/MG-Flame-report.pdf

[74] Santilli, R. M.: The novel magnecular species of hydrogen and oxygen with increased specific weight and energy content, Intern. J. Hydrogen Energy 28, 177 (2003), http://www.santilli-foundation.org/docs/Santilli-38.pdf

[75] Day, R.: TCD analysis and density measurements of Santilli Magnehydrogen. Eprida Laboratory report dated 11/10/2011 http://www.santilli-foundation.org/docs/Eprida-MH-Certification-10-11.pdf

[76] Zodape, S. P.: The MagneHydrogen in Hadronic Chemistry, AIP Proceedings 1558, 648 (2013), http://www.santilli-foundation.org/docs/sangesh-Greece.pdf

[77] Yang Y., Kadeisvili J. V. and Marton, S.: Experimental Confirmations of the New Chemical Species of Santilli MagneHydrogen, *International Journal Hydrogen Energy* 38, 5002 (2013), http://www.santilli-foundation.org/docs/Magne Hydrogen-2012.pdf

[78] Santilli, R. M.: A new gaseous and combustible form of water, *Intern. J. Hydrogen Energy* 31, 1113 (2006), http://www.santilli-foundation.org/docs/Santilli-138.pdf

[79] Storms, E.: *The science of low energy nuclear reactions* Bol. (2022), https://www.bol.com/nl/nl/f/the-science-of-low-energy-nuclear-reaction/36040739/

[80] Holgate, S. A.: *Nuclear Fusion: the Race to Build a Mini Sun on Earth* (2023), http://www.sharonannholgate.com/books/nuclear-fusion-the-race-to-build-a-mini-sun- on-earth/

[81] Santilli, R. M.: The novel Intermediate Controlled Nuclear Fusions, a report for its industrial realization. *Hadronic Journal*, 31, 15-42 (2008), http://www.santilli-foundation.org/docs/ CNF-HJ.pdf

[82] Santilli, R. M.: Experimental Confirmation of Nitrogen Synthesis from Deuterium and Carbon without harmful radiations. *New Advances in Physics*, 4, 17-41, (2010) http://www.santilli-foundation.org/docs/Nitrogen-synthesis-2010.pdf

[83] Santilli, R. M.: Additional Confirmation of the Intermediate Controlled Nuclear Fusions without harmful radiation or waste.

Proceedings of the Third International Conference on the Lie-Admissible Treatment of Irreversible Processes, Kathmandu University, 163-177, (2010), http://www.santilli-foundation.org/docs/ICNF-3.pdf

[84] Santilli, R. M.: Intermediate Controlled Nuclear Fusions without the emission of radiations and the release of radioactive waste, lecture, http://www.world-lecture-series.org/level-v

[85] Santilli, R. M.: Third Hadronic Reactor for Intermediate Controlled Nuclear Fusions without Radiations, lecture, http://www.world-lecture-series.org/lecture-vc

[86] Santilli, R. M.: Video presentation of the third hadronic reactor for the Nitrogen and Silicon syntheses, (2011), http://www.world-lecture-series.org/dragon-iii

[87] Brenna, R., Kuliczkowski T. and Ying, L.: Verification of Santilli intermediate Controlled Nuclear Fusions without harmful radiations and the production of magnecular clusters. *New Advances in Physics*, 5, 9-18, (2011) http://www.santilli-foundation.org/docs/ICNF-2.pdf

[88] Ying, L.: Professor Santilli's Intermediate Controlled Nuclear Fusions, *Proceedings of the Third International Conference on the Lie-Admissible Treatment of Irreversible Processes,* Kathmandu University, Nepal, April 2011, http://www.santilli-foundation.org/docs/Leong-Ying-2011.pdf

[89] Ying L., Cai W., J., Lynch C., Marton, Elliot S. and Yang Y.: Experimental verification for Intermediate Controlled Nuclear Fusion. City College of New York, Preprint 2012, http://www.santilli-foundation.org/docs/ICNF-Cai-paper-Ying.pdf

[90] Ying, L.: Verification of Santilli's Intermediate Nuclear Harmful Radiation and the Production of Magnecular Clusters, 2012, lecture, http://www.world-lecture-series.org/lecture-vd

[91] Kadeisvili, J. V., Lynch, C. and Yang, Y.: Confirmation of Santilli's intermediate controlled nuclear fusion of Deuterium and Carbon into Nitrogen without harmful radiation, *The Open Physical Chemistry Journal*, 5, 17-27, (2013) http://www.santilli-foundation.org/docs/ICNF-Conf-2013.pdf

[92] Lanjewar, R. B.: A Brief Review of Intermediate Controlled Nuclear Syntheses (ICNS) without Harmful Radiations. AIP Conference Proceedings, 1648, 510-515, (2015), 10.1063/1.4912717. http://www.santilli-foundation.org/docs/1.4912717(RB-Lanjewar).pdf

[93] Das Sarma, I. B.: Hadronic Nuclear Energy: An Approach Towards Green Energy. *American Journal of Modern Physics*, 5, 119-130, (2016), http://www.santilli-foundation.org/docs/pdf6.pdf

[94] Abundo, U.: Interpretation and enhancement of the excess energy of Rossi's reactor via Santilli neutroids and nucleoids. *Hadronic J.*, 37, 697-737, (2014), http://www.santilli-foundation.org/docs/abundo-paper-2014.pdf

[95] Rossiter, D., Director, IVA Report 184727-001 on comparative Nitrogen counts on samples of the Nitrogen synthesis, http://www.santilli-foundation.org/docs/nitrogen-tests-2010.pdf

[96] Rossiter, D., Director, IVA Report 200010 on comparative Nitrogen counts, http://www.santilli-foundation.org/docs/Oneida-analyses-2013.zip

[97] Rossiter, D., Director, IVA Report 184033-001 on comparative hydrogen and magnegas tests, http://www.santilli-foundation.org/docs/hydrogen-magnegas-2010.pdf

[98] Rossiter, D., Director, IVA Report 189920 on comparative Oxygen counts, http://www.santilli-foundation.org/docs/helium-carbon-fusion-I.pdf

[99] Rossiter, D., Director, IVA Report 189920 on comparative Oxygen counts, http://www.santilli-foundation.org/docs/helium-carbon-fusion-II.pdf

[100] Rossiter, D., Director, IVA Report 189920 on comparative Silicon counts, http://www.santilli-foundation.org/docs/IVAReport.189920.pdf

[101] Rossiter, D., Director, IVA Report 189920 on comparative Silicon count, http://www.santilli-foundation.org/docs/IVAReport.189920.pdf

[102] Swartz, D.: Constellation Technologies first report on comparative Silicon counts, http://www.santilli-foundation.org/docs/Constellation-Si-10-13.zip

[103] Swartz, D.: Constellation Technologies second report on comparative Silicon counts, http://www.santilli-foundation.org/docs/Constellation-Rep-Si-2.zip

[104] Swartz, D.: Constellation Technologies third report on comparative Silicon counts, http://www.santilli-foundation.org/docs/Constell-Si-3.pdf

[105] Swartz, D.: Constellation technologies Third report on comparative Silicon counts, http://www.santilli-foundation.org/docs/Constell-Silicon-10-14.pdf

[106] KAERI Table of Nuclide, website. pripyat.mit.edu/KAERI/

[107] Kashchenko, M., Kovalenko, M., Pechorsry, V. and Kashchenko, N.: Experimental confirmation of Santilli's pseudo-nuclei via the

detection of a number of modified Tungsten isotopes, in press (2025), Link to be provided.

[108] Santilli, R. M.: The Novel Hyper Combustion for the Complete Combustion of Fossil Fuels, *Intern. Journal of Chemical Engineering and Applications* 10, 16-31 (2019), http://www.santilli-foundation.org/docs/hypercombustion-2019.pdf

[109] Anderson, R.: *Collected OA Papers on Hadronic Mathematics, Mechanics, and Chemistry,* https://www.santilli-foundation.org/docs/HMMC.pdf

[110] Santilli, R.M.: An introduction to the new technologies predicted via the completion of quantum into hadronic mechanics, Invited 30' lecture at Materials 2.0, Cambridge, UK, April 27–29 ,2025, English version www.world-lecture-series.org/level-xviii/e

[111] Chinese version www.world-lecture-series.org/level-xviii/chinese Russian version www.world-lecture-series.org/level-xviii/materials2-0-ru Power point www.santilli-foundation.org/Cambridge-2025-power-point.pptx

CHAPTER 9

EXECUTIVE SUMMARY

This book is essentially a review of the lifelong mathematical theoretical, experimental and industrial research by the author [8-109] on the historical 1935 argument by A. Einstein, B. Podolsky and N. Rosen (EPR) according to which "Quantum mechanics is not a complete theory "[1-1].

The studies initiated in July 1965 with the defense of the author's thesis for his Ph. D. in theoretical physics at the University of Torino, Italy [3-1]-[3-3] on the lack of completeness of quantum mechanics for the representation of nuclear fusions. The main argument is that, in view of its anti-symmetric Lie algebra structure, Heisenberg's time evolution for an observable in terms of the Hamiltonian, $i(dA/dt) = [A, H] = AH - HA$ can only represent time reversal invariant systems with conserved total energy, $i(dH/dt) = [H,H] \equiv 0$, as it is the case for stable atoms, while energy releasing processes at large are non-conservative and irreversible over time.

To resolve the insufficiency, the Ph. D. thesis proposed the first known irreversible completion of quantum mechanics based on the generalization of Heisenberg's equation with a covering Lie-admissible and Jordan-admissible algebraic structure $i(dA/dt) = (A,H) = A<H-H>A = ARH-HSA = (ATH-HTA) + (AJH+HJA)$, $R = T+J$, $S = T-H$, $R>0$, $S>0$, $R \neq S$ [3-3] whose time rate of variation of the energy $i(dH/dt) = H(R - S)H \neq 0$ can consistently represent the energy produced by nuclear fusions.

In fact, for $R = 1$, $S = 1 - F/H$, the energy increase can be represented in terms of its analytic origin, the external terms F in Lagrange's and Hamilton's equations, $i(dH/dt) = H F$. Therefore, the proposed irreversible completion of quantum mechanics can be achieved by adding Jordan brackets to quantum mechanical Lie brackets [8-33], by therefore initiating new studies on the compatibility of mechanics and thermodynamics [3-20].

In early 1967, while holding the chair of nuclear physics at the Avogadro Institute in Torino, Italy, the indicated studies on the irreversible completion of quantum mechanics attracted an invitation to the author for a one year research contract under NASA support at the Center for Theoretical Physics of the University of Miami, Florida, as a result of which the author emigrated to the U.S.A., with his wife Carla and their two months old daughter Luisa.

In 1968, there was no mathematician or physicist in the U.S.A. with a knowledge of Lie-admissible and Jordan admissible algebras. Consequently, the author accepted a teaching position at the Department of Physics of Boston University under research support by the U. S. Air Force, where he remained as an Associate Professor of Physics for the period 1968-1974 followed by a three year stay at the Center for Theoretical Physics of MIT.

On September 8, 1977, the day of the author's arrival at Harvard University for a visiting stay, the author's supervisor received an invitation for Harvard University to apply for a research grant in support of the Santilli's research on the irreversible completion of quantum mechanics and its application in nuclear physics. As privately indicated later on by David Peaslee, the DOE Officer in charge of the grant, the unusual invitation for a grant of this advanced character apparently originated from President Jimmy Carter, who being a nuclear physicist, was technically aware of the seminal advances expected from a completion of quantum mechanics to represent the irreversibility of nature.

For an orderly and effective conduction of research under the DOE grant, the author proposed to David Peaslee:

RESEARCH STEP 1: Initiation of the research with the exact representation of all experimental data of the most fundamental fusion in nature, that of the neutron from the proton and the electron in the core of stars, which representation was missing in 1977.

RESEARCH STEP 2: Use of the methods gained in Research Step 1 to achieve an exact representation of all experimental data of the Deuteron as a bound state of a proton and a neutron, which representation was also missing in 1977.

RESEARCH STEP 3: Application of the methods gained in the structure of the neutron and of the Deuteron to the study of sustainable and controllable nuclear fusions predicted by an irreversible completion of quantum mechanics.

David Peaslee accepted the proposed research steps, and in view of the prior identification of the needed mathematics, the grant was initiated with author's joint appointment at the Lyman Laboratory of Physics and at the Department of Mathematics at Harvard University.

The official research support by the U. S. Department of Energy prompted the author and his associates around the world to plunge into intense and diversified studies on the EPR completion of quantum mechanics into a new theory under the name of hadronic mechanics, (first proposed in the 1978 memoir [3-5] and then studied in detail in the 1983 monograph [3-9] as well as in subsequent volumes), which name indicates from its Greek meaning the primary application of the new mechanics to strong interactions.

Following a preliminary study on the irreversible Lie-admissible completion of conventional reversible space time Lie symmetries and relativities presented in the 1978 memoir [3-4], the research under the DOE grant suggested the study of the transition from the historical quantum mechanical representation of dimensionless point-like particles in vacuum, into a time invariant representation of protons and neutron as they are in the physical reality, that is, extended charge distributions with an experimentally measured radius $R = 0.87 \times 10-13$ cm. It should be noted that a mathematics for the time invariant representation of extended particles did not exist in 1978, and therefore, it had to be built.

The admission of the extended character of nuclear constituents implied the admission of nuclear forces not derivable from a potential, thus not being representable with a Hamiltonian, which forces are inevitable from the experimental evidence that extended protons and neutron in a nuclear structure are in conditions of partial mutual penetration of their charge distributions (Figure 1-3), resulting in nuclear forces composed by the conventional Hamiltonian forces plus the most general known forces which are non-linear (in the wave function), non-local (because extended over the overlapping volume) and non- derivable from a potential (because of contact, thus zero range character), by therefore being beyond any hope of consistent representation by a Hamiltonian. It should be noted that in 1977 a consistent operator representation of contact non-Hamiltonian forces did not exist, and therefore, had to be found.

An understanding of the technical difficulties for the implementation of the proposed completions of quantum mechanics requires the knowledge that said completions had to be implemented by maintaining the abstract axioms of quantum mechanics in view of insurmountable technical difficulties in their surpassing (Chapter 5), as well as for the new mechanics to qualify as an EPR "completion" (rather than a "generalization") of quantum mechanics, by therefore leaving no other approach then the use of the broadest possible realizations of abstract quantum axioms.

In summary, hadronic mechanics was conceived to implement in due time the following three fundamental completions of quantum mechanics:

(1) Completion of reversible into irreversible formulations;
(2) Completion of point-like into extended particles;
(3) Completion of potential strong interactions with a non-potential component;

all completions to be implemented via broader realizations of the abstract axioms of quantum mechanics in such a way that hadronic and quantum mechanics should coincide at the abstract, realization- free level and the same should hold for specific methods, such as number theory, functional analysis, differential calculus, symmetries, relativities, etc.

The DOE supported studies from 1978 to 1983 along the above lines are reported and documented in detail in Chapter 3 [3-4]-[3-28] and include:

(I) The 1978, first known, systematic study of non-potential interactions via the conditions of variational selfadjointness (SA) [3-8] by therefore identifying the contact interactions of extended particles/wave packets as being variationally nonselfadjoint (NSA).

(II) On axiomatic grounds, quantum mechanics is based on the universal enveloping associative algebra U of operators A, B, ... with the millenary product AB verifying the associative axiom $A(BC) = (AB)C$ and related, equally millenary unit 1 verifying the axiom $1A = A1 \equiv A \; \forall A \in U$. Santilli proposed in 1978 [3-5] that hadronic mechanics should be based on the universal enveloping algebra $U*$ of the same operators A,B,... (because representing measurable quantities) equipped with the new product, called from its Greek meaning isoproduct, $A \star B = ASB$ maintaining the associative axiom $A \star (B \star C) = (A \star B) \star C$ with related generalized unit, called isounit, $1* = 1/S$ maintaining the original axiom, $1* \star A = A \star 1* \equiv A \; \forall A \in U*$, where the dimension- less quantity S, nowadays called the Santillian [6-1] [6-2] (which is fixed for a given system like the Hamiltonian) is solely restricted by the condition of being invertible, while having a completely unrestricted functional dependence on all needed local variables. Consequently, the Santillian S can indeed represent NSA forces.

The evident axiomatic equivalence of the two envelopes $U \approx U*$ assures that quantum and hadronic mechanics coincide at the abstract realization-free level, as desired, to such an extent that they could be expressed in terms of the same equations and then differentiated in their realizations. Note that this abstract identity holds despite the fact that quantum mechanics is a purely Hamiltonian theory for point-like particles under SA interactions, while hadronic mechanics is not a strictly Hamiltonian theory to represent extended particles with ensuing NSA interactions.

(III) The 1983 axiom-preserving formulation of all branches of Lie's theory, including symmetries and conservation laws, in terms of the isobrackets $[A, B]* = A \star B - B \star A = ASB - BSA$ and related isounit $1* = 1/S > 0$ [3.9], today known as the Lie-Santilli isotheory [4-12].

(IV) The 1983 identification of the Lie-isotopic dynamical equations of hadronic mechanics for stable (thus, reversible) nuclei as particular cases of the Lie-admissible equations for $R = S > 0$, which dynamical equations are given by Heisenberg-Santilli isoequation with a Lie-isotopic structure $i(dA/dt) = [A.H]* = A \star H - H \star A = ASH - HSA$ and Schrödinger-Santilli isoeigenvalue equation $H \star |\psi\rangle = HS|\psi\rangle = E|\psi\rangle$, where the Hamiltonian H represents all potential interactions and the Santillian S represents the extended character of particles and their contact interactions [3-9].

(V) The 1983 axiom-preserving Lie-isotopic completion of Galileo's relativity [3-9] and Einstein's special relativity [7-5] which are necessary for the corresponding non- relativistic and relativistic invariance over time of the hadronic mechanics, that is, the prediction of the same numeric values under the same conditions at different times which is essential for credible experimental verifications.

(VI) The 1981 admission of causal superluminal (e.g., tangential) speeds under contact non-Hamiltonian interactions due to their proved capability of accelerating particles without the use of energy [3-24] [8-31] [8-32].

(VII) The biggest technical difficulty for Research Step 1 is provided by Heisenberg's uncertainty principle because it prohibits any representation of the synthesis of the neutron as a bound state of the proton and the electron despite their Coulomb attraction due to their opposing charges which reaches the simply astronomical value (for particle standards) of 230 Newton at the charge radius of the neutron. In the seminal, Harvard University, DOE supported and recorded paper [3-21], Santilli: (a) Noted that Heisenberg's uncertainty principle has been experimentally proved only for atomic electrons under long range electromagnetic interactions and merely assumed to be valid within nuclear structures without any experimental verification; (b) Argued that, in view of the known strength of nuclear forces compared to electromagnetic atomic forces (Figure 6.1), the uncertainties within nuclei are expected to be much weaker than the corresponding uncertainties within atoms; (c) Proposed the progressive recovering of Einstein's determinism [1-1] with the increase of nuclear densities $\Delta r \Delta p = (1/2)|\langle\psi| \star [r, p]* \star |\psi\rangle| \rightarrow 0$.

(VIII) The 1979 axiomatic compatibility of the irreversible Lie-admissible branch of hadronic mechanics with the abstract axioms of reversible quantum mechanics was achieved via the decomposition of the enveloping isoassociative algebra U∗ of hadronic mechanics into the product of two ordered isoassociative algebras, U∗ ≈< U×U>, the first isoalgebra <U is assumed to have all ordered products to the left A < B = ARB and to represent motion backward in time while the second isoalgebra U> is assumed to have all ordered products to the right A > B = ASB, R≠ S and to represent motion forward in time. This important axiomatic unification of the irreversibility of hadronic mechanics with the reversibility of quantum mechanics was first proposed by Santilli in the DOE supported paper [3-19] released under the author's affiliation with the Department of Mathematics of Harvard University.

(IX) The 1978 representation of all characteristics of mesons and leptons via new hadronic structure models with physical constituents produced free in the spontaneous decays. The new models were presented in Table 5.1 of memoir [3-5] to illustrate the capabilities permitted by hadronic mechanics while being prohibited by quantum mechanics. As an example, Santilli showed that the representation of the π0 meson as a bound state of an electron and a positron at 10−13 cm mutual distances according to hadronic mechanics (hm), i.e. as a "compressed positronium", π0 = (e− ↑, e+ ↓)hm permits the numerically exact and time invariant representation of all characteristics of the π0 (Eqs. (5.1.14), page 886 of memoir [3-5]), including the first and only known identification of the mechanism triggering the spontaneous decay as being due to electron-positron annihilation. Intriguingly, the strong interactions emerging in the compression of the positronium are represented with the initial simple Santillian S = exp(−Vh/Vc) ≈ 1 − Vh/Vc which replaces the Coulomb potential Vc with the much stronger Hulten potential Vh in the

Schrödinger-Santilli isoequation $H \star |\psi\rangle = HS|\psi\rangle = E|\psi\rangle$. Similar representations of all characteristics via hadronic mechanics hold for the remaining mesons and for the leptons known in 1978.

The studies from 1984 to today (July 9, 2025) in the construction of hadronic mechanics gave priority to the development of its new isomathematics which reached maturity for applications thanks to the following primary advances:

(X) Due to its enveloping associative algebra with product AB, 20th century applied mathematics is formulated over numeric fields F with the number 1 as the multiplicative unit. Since isomathematics is based on the isoassociative product $A \star B = ASB$, in the 1993 paper at Algebras, Groups and Geometries [4-1] Santilli extended the number theory to hold for arbitrary invertible isounits $1* = 1/S$, resulting in the consistent formulation of hadronic mechanics over numeric isofields $F*$ with isonumbers $n* = n1*$, where n rep- resents ordinary real, complex and quaternionic numbers.

(XI) Despite the consistent formulation over isofields, in the early 1990's hadronic mechanics had not yet reached the crucial invariance over time of numeric predictions. Following extensive studies, Santilli identified the problem in the lack of compatibility of completions (1) (2) (3) and the Newton-Leibnitz differential calculus due to its sole possible definition at isolated points and achieved in the 1996 Special Issue of the Rendiconti del Circolo Matematico di Palermo [4-2] the needed broader differential calculus with isounits dependent on the differentiation variable, such as time t, coordinates r, velocities v, etc., $1* = 1*(t, r, v, ...)$ which broader calculus can be summarized as follows. Within the context of isomathematics all quantities with numeric values must be isonumbers $n* = n1*$, i.e., must be elements of an isofield and this includes isocoordinates $r* = r1*$ and isovelocities $v* = v1*$. Consequently, by assuming the general form of the isounit $1* = \exp\Gamma(t, r, v, ...) = 1/S > 0$, Santilli introduced the isotopic generalization of Newton's differential $d*v* = S\, d[v1*(t,...)] = dv + v\, S\, d1*(t,,r,,v,...) = dr + v\, d\Gamma(t, r, v,...)$ which embeds Newton's historical, velocity dependent resistive (well as all NSA) forces in the isotopic generalization of the differential and of related derivative. The reader should be aware that most of novel physical results presented in this book are due to the extra term $vd\Gamma$ appearing in Santilli's isodifferential.

(XII) Maturity of formulation of isomathematics was achieved in the late 1990's thanks to the following monographs, each one having extended references, authored by senior mathematicians, such as: the Greek mathematicians D. S. Sourlas and Gr. Tsagas who studied the isotopies of Lie theory [4-12]; the Chinese mathematician C-X Jiang who studied the isonumber theory [4-14]; the Spanish mathematicians R. M. F. Ganfornina and J. N. Valdes who studied the isotopies of quantum mechanics [4-15]; the French mathematician S. Georgiev who conducted truly extensive studies on the isodifferential calculus [4-17]; and others.

Following the achievement of maturity in the formulation of isomathematics, the Lie- isotopic branch of hadronic mechanics reached maturity for experimental verifications and industrial applications thanks to a number of advances in basic aspects, among which we note:

(XIII) Heisenberg's, Schroedinger and Dirac's equations resulted to be inapplicable for a quantitative treatment of Research Step 1 (the exact and invariant representation of the neutron synthesis from the proton and the electron) in favor of Lie-isotopic dynamical equations [8-31] because the rest energy of the neutron is 0.782 MeV bigger than the sum of the rest energies of the proton and of the electron and for other technical reasons.

(XIV) The same quantum equations resulted to be inapplicable for Research Step 2 (the exact and invariant representation of the experimental data of the Deuteron and other light nuclei) in view of: their only possible proton-neutron bound state in singlet coupling with spin 0 in the ground state contrary to the spin1 of the Deuteron their approximation of protons and neutrons as massive points in vacuum, with consequential inability to represent the mutations of the intrinsic characteristics of extended charge distributions under strong interactions [3-5] which are prohibited by 20th century space time symmetries and relativities but readily admitted by the covering isotopic symmetries and relativities (Chapter 7); and for other reasons.

(XV) Quantum mechanics resulted to be unable of representing particle entanglements in favor of the hadronic entanglement [6-7]. This is due to the lack of any known quantum mechanical potential for large mutual distances, while particles have wave packets that extend to infinity and the contact non-potential interactions in their overlapping persists at arbitrary distances (Figure 6.3).

(XVI) Bell's inequalities on the lack of classical counterparts for systems of point like protons and neutrons with spin 1/2 resulted to be inapplicable for extended protons and neutrons with spin 1/2 in favor of hadronic inequalities which admit full classical counter parts [6-9], by therefore providing the first opening for the progressive recovering of Einstein's determinism.

(XVII) Hadronic mechanics was first known to provide explicit and concrete realizations of Bohm's "hidden variables" λ in page 254 of monograph [4-5] as being hidden in the isoassociativity axiom $A \star B = A\lambda B$ with their ensuing concrete realization by the Santillian $\lambda = S$, resulting in: a clear opening for recovering Einstein's determinism; the exact and invariant representation of the Deuteron magnetic moment [8-38] [8-39]; and other advances.

(XVIII) Following the 1979 DOE recorded conception of the hadronic uncertainties [3-21], and the 1994 proof of the full recovering by hadronic mechanics of Einstein's determinism for gravitational collapse [6-14], Santilli finalized the uncertainties of hadronic mechanics under the name of Einstein's isodeterminism in the 2019 paper published by Ratio Mathematica [6-15] $\Delta r \Delta p = (1/2)\ |\langle\psi| \star [r, p]* \star |\psi\rangle| \approx (1/2)\ S \to 0$ which was formalized via a theorem in the 2020 paper [6-17] essentially representing the progressive recovering of Einstein's determinism with the increase of the density in the interior of hadrons, nu- clei and stars and its full recovering at the limit of Schwartzschild's horizon (Chapter 6). Intriguingly, a Corollary of the above theorem (page 128 of [6-17]) states that the progressive recovering of Einstein's determinism implies the removal of the divergencies in particle physics because the insertion of the sufficiently small Santillian S in between all products $AB \to A \star B = ASB$ implies the conversion of strongly divergent quantum series into strongly convergent isoseries.

(XIX) Planck's quantum theory resulted to be inapplicable in nuclear physics due to the lack of excited nuclear orbits, the lack of quantized nuclear binding energies, the lack of quantized energy emission in beta decays, et al., in favor of Santilli's integro-differential isounit $1* = 1/S > 0$. The main difference between quantum and hadronic mechanics is therefore given by discrete energy exchanges for quantum mechanics compared to non- local energy exchanges for hadronic mechanics [4-1].

(XX) Inapplicability for extended protons and neutrons of Galileo's and Einstein's relativities in favor of their axiom-preserving hadronic completions, resulting in a major broadening (rather than violation) of their representational capabilities, including: the experimentally established local character of the frequency and speed of electromagnetic waves within physical media; causal superluminal speeds (e.g., tangential) within nuclear media triggered by Santilli's interactions due to their proved acceleration of particles without the use of energy; and other advances (Chapter 7).

Nowadays, hadronic mechanics and its new mathematics are seen as the new paradigm for the representation of Einstein's spooky action at a distance, namely, the invisible, continuous and instantaneous interactions between entangled particles [1-1] caused by the non-Hamiltonian overlapping of the wave packets of particles [6-7] that, in view of its existence also at a classical level [6-21], is having basic applications in various branches of physics, chemistry and biology (see the recent general review by E. Velardo [1-21]).

We now outline the various applications of hadronic mechanics to nuclear physics presented in Chapter 8 which have been done via the use of the entirety of the methods presented in Chapters 1 to 7 with realizations of the Santillian for a system of two entangled particles of the type S = Σα=1,2 Diag(1/n2α,1, 1/n2α,2, 1/n2α,3, 1/n2α,4) exp(−Γ) > 0, where: n2α,k , k = 1, 2, 3, represents the semi axes of the particles α = 1, 2 normalized to the value n2α,k = 1 for the circle; n2α,4 represents the density of particles α = 1, 2 normalized to the value n2α,4 = 1 for the vacuum and the quantity Γ > 0 represents the non-Hamiltonian interactions due to the overlapping/entanglement of the wave packets of particles α = 1,2 with an unrestricted functional dependence on time t, coordinates r, velocities v, acceleration a, energy E, wave function ψ as well as density d, temperature τ and pressure π of the surrounding medium, Γ = Γ(t, r, v, a, E, ψ, d, τ, π , ...) > 0:

9.1. Insufficiencies of quantum mechanics in nuclear physics

Quantum mechanics has provided extraordinary insights into atomic systems but has proven inadequate for nuclear physics. This section reaffirms the various key limitations of quantum mechanics in favor of hadronic mechanics.

9.2. The synthesis of neutrons and neutroids from the Hydrogen

Hadronic mechanics resolves the neutron synthesis problem by introducing isoneutrons, hadronic bound states formed under conditions of total penetration of the electron within the hyperdense proton under the immense pressures in the core of stars. The model accounts for all neutron properties - including mass, charge radius, spin, magnetic moment, mean life and spontaneous decay - without violating energy conservation. This synthesis, originally envisioned by Rutherford in 1920, is now rendered consistent via isomathematics. The Don Borghi neutroid, formed in low-energy conditions, is a precursor to the isoneutron. The relativistic hadronic model of the neutron synthesis explains the vastly ignored gravitational stability of the Sun. The theory proposes a shift from neutrino-based models to etherino-based transitions while representing existing particle experiments, suggesting an alternative to the neutrino hypothesis.

9.3. The syntheses of protoids and pseudo-protons

Hadronic mechanics extends its predictive power to introduce two new nuclear constituents: the protoids, synthesized under high-energy arc discharges, which are new particles intermediates between hydrogen and nuclear matter and the pseudo-protons, both predicted to exist in the Sun following the neutron synthesis via deep electron-neutron penetrations, and both being attracted by ordinary nuclei, making them ideal for initiating novel nuclear reactions. These particles provide new mechanisms for the recycling of nuclear waste and for triggering fusions without the Coulomb barriers. Experimental evidence for their synthesis has been provided by U. S. spectroscopic laboratories and by the DeHaan pseudo-protonium experiments.

9.4. Representation of the Deuteron and pseudo-Deuteron

The structure of the Deuteron D, long problematic under quantum mechanics, is clarified within hadronic mechanics: Hadronic models reinterpret the Deuteron as a hadronic bound state of a proton and a neutron, and ultimately, of an isoproton and an isoproton and an isoneutron thanks to the inclusion of contact interactions. Additionally, thanks to Einstein's isodeterminism, hadronic mechanics has permitted the experimentally confirmed synthesis of negatively charged pseudo-nuclei and, in particular, of the pseudo-Deuteron D^* with negative charge -1 as a hadronic bound state of an electron pair (ee) in singlet coupling and a natural Deuteron with ensuing natural fusion into the Helium $D + D^* \rightarrow He + 2e$ without the Coulomb barrier, without the emission of harmful radiation and without the release of radioactive waste.

9.5. Experimental confirmation of neutrons and neutroid syntheses

This section presents direct experimental evidence in support of hadronic predictions: Don Borghi's experiments established the existence of neutroids via low-energy reactions in Hydrogen gas. The author and his technicians reproduced in laboratory the neutron synthesis in the core of stars via electric arcs submerged in Hydrogen, confirming the neutron's hadronic model. The commercially available Directional Neutron Source, that was developed using these principles, has achieved consistent and industrially usable low energy directional neutron beams.

Further confirmation comes from: the synthesis of pseudo-protonium atoms by Dr. V. de Haan, observed via their anomalous spectral and structural features; the directional emissions of low energy neutrons that are impossible to explain using conventional nuclear models, but naturally arise in hadronic treatments of wave packet overlaps.

9.6. The recycling of radioactive nuclear waste according to hadronic mechanics

Hadronic mechanics allows for three distinct mechanisms of radioactive waste recycling via irradiations, each physically realizable by nuclear power plants:

(A) Low energy neutron irradiation – e. g., via the author's DNS) - which activates nuclear transmutations with a bigger atomic number and consequential rapid decay while preserving the protonic number.

(B) Stimulated decay of peripheral neutrons triggered by high energy photons with a special resonating frequency which with increased protonic number while preserving the atomic number and consequential rapid decay.

(C) Pseudo-proton irradiations to trigger transmutation into bigger atomic and protonic numbers, which irradiation is moist effective due to the negative charge of the pseudo-protons and their consequential Coulomb attraction by the nuclei of the radioactive waste produced by nuclear power plants.

All methods are permitted within hadronic mechanics and have been partially validated through experimental transmutations of natural stable elements. Their implementation can be realized within existing nuclear power plants, removing the need for transportation and long-term waste storage.

9.7. Clean nuclear energies according to hadronic mechanics

This pivotal section lays out the theory and engineering of Intermediate Controlled Nuclear Fusions (ICNF), a novel class of clean fusion reactions based on hadronic mechanics.

9.7.1. Clean nuclear energies via stimulated beta decay

Stimulated beta decay is employed to produce clean fusion energy. This process relies on high-energy photons inducing neutron decays within hadronic bound states with ensuing nuclear transmutations, bypassing quantum-mechanical limitations.

9.7.2. Tsagas confirmation of Santilli's stimulated beta decay

The Greek nuclear physicist Gr. Tsagas independently confirmed this mechanism by measuring radiation-free transmutations in beta-decaying isotopes irradiated with specific photons.

9.7.3. Clean nuclear energies via pseudo-proton Irradiations

Irradiating materials with pseudo-protons-synthesized within hadronic reactors has led to transmutations that release measurable excess heat without harmful emissions. These negatively charged strongly interacting particles are attracted by positively charged natural nuclei by therefore bypassing the Coulomb barrier.

9.7.4. Physical laws of controlled nuclear fusions according to hadronic mechanics

Santilli outlines five foundational laws derived from extensive tests in the field to enhance sustainable Intermediate Controlled Nuclear Fusions (ICNF).

9.7.5. The new chemical species of magnecules.

To enhance ICNF, Santilli has discovered and developed the new chemical species of magnecules - clusters of toroidal polarized atoms held by magnetic bonds - for the systematic and controlled exposure of nuclei out of their electron clouds, including:

Magnegas (MG), which is a Hydrogen-based, clean-burning fuel with high flame temperature;

MagneHydrogen (MH), which is a magnecular species of Hydrogen with increased specific density due to magnetic bonds of polarized H-atoms, and consequently, increased energy content per mole and without seepage when stored in tanks at pressure;

HHO, the new gaseous form of water requiring no Oxygen for its combustion in air or underwater and with an energy content so anomalous to cut bricks at contact.

9.7.6. Intermediate Controlled Nuclear Fusions (ICNF)

Hadronic reactors using gaseous fuels with submerged electric arcs can reproducibly initiate ICNF. Subsections detail for specific fuel combinations:

9.7.6-I: ICNF with the Nitrogen hadronic reactor; **9.7.6-II:** ICNF with Deuterium and Carbon; **9.7.6-III:** ICNF with Hydrogen and Carbon; **9.7.6-IV:** ICNF with Magnegas and Carbon; **9.7.6-V:** ICNF with the Oxygen hadronic reactor; **9.7.6-VI:** ICNF with the Silicon hadronic reactor.

9.7.7. Independent certifications of clean excess energy produced by ICNF hadronic reactors

Multiple independent groups have confirmed the production of clean excess thermal energy from ICNF reactors, including:

9.7.7-I: Princeton Gamma Tech Instruments detected anomalous heat consistent with non-chemical reactions;

9.7.7-II: Y. Yang et al. performed spectral and isotopic analyses confirming pseudo-nuclear formation and transmutation;

9.7.7-III: Additional laboratory verifications supported the existence of new isotopic species and the presence of clean excess energy.

These findings validate ICNF as a safe, clean, and scalable energy generation process.

9.8. HyperFusions (HF)

HyperFusions represent the apex of hadronic fusion technology. They are ICNF between negatively charged light pseudo-nuclei and positively charged natural light nuclei, thus occurring without the Coulomb barrier.

9.8.1. The birth of HyperFusions

Born from ICNF, HF reactions exploit deeper penetration and higher-density wave packet overlap. This results in reactions without the emission of harmful radiation and without the release of radioactive waste.

9.8.2. Insufficiency of calorimetric measurements

Traditional calorimetry cannot reliably capture the excess energy outputs of HF although they can only measure part of the energy output of HF in favor of the measurement of the energy produced by the collection of the spectroscopically detected actual nuclear fusions.

9.8.3. Experimental confirmation of pseudo-nuclei by the Ural Federal University, Yekaterinburg, Russia

At the Ural Federal University, researchers conducted electric arc experiments by using Tungsten as electrodes. Post-reaction analysis detected the increase of the mass by a number of Tungsten isotope, by therefore providing experimental confirmations of the lack of validity of Heisenberg's uncertainty principle in nuclear physics, as well as the prediction by hadronic mechanics of bonds between electrons and natural nuclei.

9.8.4. Clean excess energy from detected nuclear fusions

In view of the indicated insufficiencies of thermal calculations of excess heat produced by hadronic reactors, in this section, we report the excess energy output presented in paper [8-33] as direct calculations of the energy produced by the following representative count per part per million per volume (ppmv) of nuclear fusions with atomic mass de- noted with the letter u (ORS laboratories [8-94]) (a)Δ 2 u: 18, 550, 801 – 16, 075, 402 =

2, 475, 399 ppmv,(b)Δ 3 u: 41, 165 — 30, 269 = 10, 896 ppmv, (c)Δ 4 u = 76 – 0 =76 ppmv, (d)Δ 14 u : 3, 555 – 2, 841 = 714 ppmv, €Δ 16 u : 3, 010 – 1, 205 = 1, 805 ppmv, (f)Δ 18 u 2, 949 – 2, 718 = 231 ppmv, (g)Δ 28 u: 30, 171 – 24, 684 =3, 687 ppmv.

By identifying the actual nuclear fusion per count, the above ORS data imply the following corresponding energy production (see Ref. [9-33] for details and the Table of Nuclides [8-105]) $E_a = 2.014102$ u, $E_b = 4.002603$ u, $E_c = 12.00000$ u, $E_d = 14.003074$ u, $E_e = 15.994915$ u, $E_f = 15.004915$ u, E(g) = 27.976927u for an approximate total in parts per millions per mole of 54, 546, 518 MeV $= 2.43 \times 10^{-12}$ kwh.

By assuming that the gas in the reactor is a perfect gas, by assuming the related law $PV = nRT$ and by recalling that one mole contains 6.02×10^{23} particles, that is, 6.02×10^{17} *millions* of particles, the total energy output calculated in paper [8-33] is given by $\Delta E = E^{out} — E^{in} = 2.43 \times 10—12 \times n \times 6.02 \times 10^{17} — 1.333$ kwh $\approx n \times 14.628 \times 10^5$kwh where the number of moles of hadronic fuel in the interior of the hadronic reactor is given by (see paper [8-33] for details) $n \approx 7 \times 10^{-5}$.

Consequently, the approximate total output of controlled and sustained clean energy of the considered ICNF is given by $\Delta E = E^{out} - E^{in} = 7 \times 10^{-5} \times 15 \times 10^5 - 1.333$ kwh ≈ 100 kwh under the operating conditions set in Section 8.7.6-VI (50 kwh for AC-DC converter, gaseous hadronic fuel set at 1000 psi and operations for 15 m), out of which 70 kwh are predicted to be lost in the conversion of heat to electricity, by remaining with 30 kwh of clean electricity produced by a modular reactor sufficient for households or small factories.

It should be finally indicated that, under the engineering conditions of Sect. 8.7.6, HyperFusions cannot produce atomic explosions because, in view of inevitable random processes in any plasma, they are naturally restricted to occur in parts per million per volume of one mole (ppmv

PROPOSED EXPERIMENTS

PROPOSED EXPERIMENT 1:
Confirm or deny the existence of Don Borghi's neutroid
Section 8.2-2

PROPOSED EXPERIMENT 2:
Confirm or deny the existence of Santilli's protoids and/or pseudo-protons
Sections 8.3-1 to 8.3-3

PROPOSED EXPERIMENT 3:
Confirm or deny the existence of de Haan's pseudo-protonium
Section 8.5-6

PROPOSED EXPERIMENT 4:
Confirm or deny the recycling of radioactive nuclear waste via neutron, pseudo-proton or other thermal irradiation
Section 9.5-4, 8.6-1 and 8.6-3

PROPOSED EXPERIMENT 5:
Confirm or deny Tsagas experiment on stimulated beta decays
Section 8.7.2

PROPOSED EXPERIMENT 6:
Confirm or deny Kashchenko's synthesis of pseudo-nuclei
Section 8.8-3

PROPOSED EXPERIMENT 7:
Confirm or deny Santilli's hyperfusions without the Coulomb barrier
Section 8.8

For technical details and possible funding, please contact
The R. M. Santilli Foundation
board(at)santilli-foundation(dot)org
https://www.santilli-foundation.org/news.html

AUTHOR INDEX

SUBJECT INDEX

AUTHOR BIOGRAPHY

Sir Prof. Ruggero Maria Santilli

Sir Santilli is an Italian-American scientist who graduated in theoretical physics from the University of Naples, Italy, in July, 1960; obtained his Ph. D. at the University of Torino, Italy, in 1965; and then emigrated to the U.S.A. in 1967 where he was on the faculty of the University of Miami, Boston University, MIT, and Harvard University under financial support from NASA, UFOSR and DOE. In 1991 he accepted the position of President and Professor at the Institute for Basic Research in Florida. He is the author of 359 papers published in refereed journals and twenty post Ph. D. monographs in mathematics, physics, chemistry and biology; he is the Editor of a number of scientific journals in mathematics and physics and was awarded twelve U. S. patents. He is internationally known for developing new mathematics (known as Santilli's Lie-admissible, Lie-isotopic and isodual mathematics) for the construction of hadronic mechanics, chemistry and biology, and for the verification of the 1935 Einstein-Podolsky-Rosen argument that "quantum mechanics is not a complete theory." These discoveries lead to the prediction of new environmental technologies, incompatible with quantum mechanics but fully compatible with hadronic mechanics, which he developed as Chief Scientist at various U. S. publicly traded and private companies. Ruggero Maria Santilli was knighted with the title of Sir on September 8, 2011, by the Republic of San Marino with the Membership in the Millenary Equestrian Order of Sant'Agata. Additionally, Sir R.M. Santilli has received: the Gold Medal from the Universite' d'Orleans, France, on January 6, 1982; the Nomination among the most illustrious applied mathematicians if all times by the Estonia Academy of Sciences in June 9, 1990; a dedication to his name of a lecture hall by the Australian Science and Research Center on September 9, 1992; the Scientific Prize of the Mediterranean Foundation on January 2, 2009; the Frey International Sustainability Award on November 9, 2016; and the "Stella d'Italia" (Star of Italy) with the title of "Cavaliere" from the Italian President Sergio Mattarella on May 31, 2018; and other important awards. For details. For more information, please visit the website http://www.i-b-r.org/Sir-Santilli-bionotes-05-15-21.pdf